Methods and Protocols in Food Science

Series Editor
Anderson S. Sant'Ana
University of Campinas
Campinas, Brazil

For further volumes:
http://www.springer.com/series/16556

The Methods and Protocols in Food Science series is devoted to the publication of research protocols and methodologies in all fields of food science.

Volumes and chapters will be organized by field and presented in such way that the readers will be able to reproduce the experiments in a step-by-step style. Each protocol will be characterized by a brief introductory section, followed by a short aims section, in which the precise purpose of the protocol will be clarified.

Frozen Dough and Bread Quality

Edited by

Guoqiang Zhang

Coventry University, Coventry, UK

Jinying Guo

College of Food and Bioengineering, Henan University of Science and Technology, Luoyang, China

Humana Press

Editors
Guoqiang Zhang
Coventry University
Coventry, UK

Jinying Guo
College of Food and Bioengineering
Henan University of Science and Technology
Luoyang, China

ISSN 2662-950X ISSN 2662-9518 (electronic)
Methods and Protocols in Food Science
ISBN 978-1-0716-5205-3 ISBN 978-1-0716-5206-0 (eBook)
https://doi.org/10.1007/978-1-0716-5206-0

This Humana imprint is published by the registered company Springer Science+Business Media, LLC, part of Springer Nature.
The registered company address is: 1 New York Plaza, New York, NY 10004, U.S.A.

If disposing of this product, please recycle the paper.

Preface to the Series

The *Methods and Protocols in Food Science* series is devoted to the publication of research protocols and methodologies in all fields of food science. The series is unique as it includes protocols developed, validated, and used by food and related scientists, and theoretical basis is provided for each protocol. Aspects related to improvements in the protocols, adaptations, and further developments in the protocols may also be approached.

The *Methods and Protocols in Food Science* series aims to bring the most recent developments in research protocols in the field as well as very well established methods. As such, the series targets undergraduate, graduate, and researchers in the field of food science and correlated areas. The protocols documented in the series will be highly useful for scientific inquiries in the field of food sciences, presented in such a way that the readers will be able to reproduce the experiments in a step-by-step style.

Each protocol will be characterized by a brief introductory section, followed by a short aims section, in which the precise purpose of the protocol is clarified. Then, an in-depth list of materials and reagents required for employing the protocol is presented, followed by a comprehensive and step-by-step procedure on how to perform that experiment. The next section brings the dos and don'ts when carrying out the protocol, followed by the main pitfalls faced and how to troubleshoot them. Finally, template results will be presented and their meaning/conclusions addressed.

The *Methods and Protocols in Food Science* series will fill an important gap, addressing a common complaint of food scientists regarding the difficulties in repeating experiments detailed in scientific papers. With this, the series has the potential to become a reference material in food science laboratories of research centers and Universities throughout the world.

Campinas, Brazil *Anderson S. Sant'Ana*

Preface

Bread is one of the most consumed daily staple foods in the world, providing significant amounts of energy and nutrients to people's daily diet. Therefore, this sector has become one of the most important positions in the global food industry. Frozen dough is widely used in the baking industry to meet the growth of product flexibility and consumers' pursuit of freshness and quality. With the development of cold chain transportation, the application of frozen dough has further promoted the development of the baking industry, and its requirement is further increasing. Notably, dough preparation is the first step in bread making; therefore, the quality of frozen dough is crucial in determining the final quality of bread. Both the food industry and researchers are dedicated to improving frozen dough quality to meet consumer demand for high-quality and nutritious bread products. Consequently, understanding the quality evaluation of frozen dough and related bread products is essential for driving further innovation and development.

In this context, this book, *Frozen Dough and Bread Quality*, aims to provide comprehensive knowledge and detailed protocols in the analysis of frozen dough and related bread products. Overall, this book covers a wide spectrum of topics within frozen dough bread, ranging from frozen dough production, related bread baking, and the quality assessment of both the dough and bread products. This book will not only provide a comprehensive guide for the growing number of researchers, professionals, and students working today in the field of bakery products but will also serve as a solid foundation for continued innovation and development in frozen dough and related bread products, helping to produce desirable quality bread that meets consumer expectations.

I would like to express my deepest gratitude to all the authors for their invaluable insights, expertise, and knowledge, as well as to the publishers for their efforts, which make this book possible.

I sincerely thank all readers for their interest in this book, and I hope that this book will serve as a valuable reference and practical guide, supporting you in finding the information you need and meeting your expectations.

Coventry, UK *Guoqiang Zhang*

Contents

Contributors

FEIYU AN • *College of Food Science, Liaoning Engineering Research Center of Food Fermentation Technology, Shenyang Key Laboratory of Microbial Fermentation Technology Innovation, Shenyang Agricultural University, Shenyang, China*

SHUQI AN • *College of Food Science, Liaoning Engineering Research Center of Food Fermentation Technology, Shenyang Key Laboratory of Microbial Fermentation Technology Innovation, Shenyang Agricultural University, Shenyang, China*

XIAOSHUANG CAI • *College of Food Science and Engineering & Institute of Special Oilseed Processing and Technology, Henan University of Technology, Zhengzhou, China*

JIALU CHEN • *College of Food and Bioengineering, Henan University of Science and Technology, Luoyang, China*

XI-CHUANG CHENG • *College of Food Science and Engineering & Institute of Special Oilseed Processing and Technology, Henan University of Technology, Zhengzhou, China*

JINBO GUO • *College of Food Science, Shanxi Normal University, Taiyuan, China*

JINYING GUO • *College of Food and Bioengineering, Henan University of Science and Technology, Luoyang, China*

XIANG JIN • *College of Food Science, Liaoning Engineering Research Center of Food Fermentation Technology, Shenyang Key Laboratory of Microbial Fermentation Technology Innovation, Shenyang Agricultural University, Shenyang, China*

ZHANGCHENG LIANG • *Institute of Food Science and Technology, Fujian Academy of Agricultural Sciences, Fuzhou, Fujian, China*

HUA-MIN LIU • *College of Food Science and Engineering & Institute of Special Oilseed Processing and Technology, Henan University of Technology, Zhengzhou, China*

YUXUAN LIU • *College of Food Science, Liaoning Engineering Research Center of Food Fermentation Technology, Shenyang Key Laboratory of Microbial Fermentation Technology Innovation, Shenyang Agricultural University, Shenyang, China*

YUANQING LUO • *College of Food Science and Engineering & Institute of Special Oilseed Processing and Technology, Henan University of Technology, Zhengzhou, China*

ZHAO QIN • *College of Food Science and Engineering & Institute of Special Oilseed Processing and Technology, Henan University of Technology, Zhengzhou, China*

LING-YU QU • *College of Food Science and Engineering & Institute of Special Oilseed Processing and Technology, Henan University of Technology, Zhengzhou, China*

YANGE TIAN • *College of Food and Bioengineering, Henan University of Science and Technology, Luoyang, China*

CHUNYAN WANG • *Food College, Shenyang Agricultural University, Shenyang, China*

SIJIA WANG • *College of Food Science and Engineering, Henan University of Technology, Zhengzhou, China*

WENMIN WANG • *College of Food Science, Liaoning Engineering Research Center of Food Fermentation Technology, Shenyang Key Laboratory of Microbial Fermentation Technology Innovation, Shenyang Agricultural University, Shenyang, China*

YUANHUI WANG • *College of Food Science and Engineering, Henan University of Technology, Zhengzhou, China*

JIA WEI • *Institute of Food Science and Technology, Xinjiang Academy of Agricultural Science, Urumqi, Xinjiang, China*
JUNRUI WU • *College of Food Science, Liaoning Engineering Research Center of Food Fermentation Technology, Shenyang Key Laboratory of Microbial Fermentation Technology Innovation, Shenyang Agricultural University, Shenyang, China*
RINA WU • *College of Food Science, Liaoning Engineering Research Center of Food Fermentation Technology, Shenyang Key Laboratory of Microbial Fermentation Technology Innovation, Shenyang Agricultural University, Shenyang, China*
WENXIN YANG • *College of Food Science, Liaoning Engineering Research Center of Food Fermentation Technology, Shenyang Key Laboratory of Microbial Fermentation Technology Innovation, Shenyang Agricultural University, Shenyang, China*
YIXI YANG • *College of Food Science and Engineering, Henan University of Technology, Zhengzhou, China*
GUOQIANG ZHANG • *School of Science, Coventry University, Coventry, UK*
QI-WEI ZHANG • *College of Food Science and Engineering & Institute of Special Oilseed Processing and Technology, Henan University of Technology, Zhengzhou, China*
RUN-YANG ZHANG • *College of Food Science and Engineering & Institute of Special Oilseed Processing and Technology, Henan University of Technology, Zhengzhou, China*

Part I

Frozen Dough Preparation and Analysis

Check for updates

Chapter 1

Frozen Dough Preparation

Guoqiang Zhang and Yuanhui Wang

Abstract

Frozen dough has been widely applied in food industry for bread production. Frozen dough is produced by directly freezing dough after preparation to preserve it for further processing. This approach not only reduces the production burden of baked products, but also flexibly meets consumer demand for high quality and freshly, offering an efficient pathway for the stable production of fresh baked goods. Based to differences in the fermentation, the frozen dough can be categorized into unfermented frozen dough and prefermented frozen dough. Dough preparation is the first step in bread production, and its processing procedure is crucial in transforming ingredients into the final product. Therefore, the aim of this chapter is to provide a clear procedure for frozen dough preparation, focusing primarily on unfermented frozen dough.

Keywords Frozen dough, Frozen dough preparation, Unfermented frozen dough, Bread formulation, Baking products

1 Introduction

Bread is one of the most widely consumed and popular bakery products globally and is an important part of the human diet [1]. The global bread market size is steadily increasing, valued at 234.43 billion USD in 2024 and projected to reach 290.80 billion USD by 2030, with a compound annual growth rate of 3.71% [2]. However, breadmaking is inherently complex, labor, and time-consuming process, especially in dough making, which is most critical step in breadmaking [3]. Meanwhile, due to the staling, the quality of freshly baked bread deteriorates rapidly during storage, adversely influencing texture and sensory characteristics of bread and further consumer acceptance [4]. To this end, bread-making technology continues to evolve to meet these demands. Frozen dough technology has received much attention and gradually become the basis for large-scale baking production [5, 6]. Frozen dough technology refers to the use of freezing technology to preserve prepared dough for a certain period of time, offering a convenient and efficient method to

Guoqiang Zhang and Jinying Guo (eds.), *Frozen Dough and Bread Quality*, Methods and Protocols in Food Science,
https://doi.org/10.1007/978-1-0716-5206-0_1,

maintain dough quality for long period and reduce the production burden of final products [7, 8]. This approach also greatly facilitates consumers to ensure a stable supply of fresh baked products, allowing them to obtain freshly baked products as needed [9].

In general, frozen bread dough typically consists of four key ingredients, including flour, water, salt, and yeast [10]. In addition, a variety of ingredients can be added into formulation for bread dough making such as milk, milk powder, eggs, sugar, spices, butter, fruits, vegetables, and other novel food ingredients (e.g., insect flours) [11–13]. Notably, different dough formulations exist in various countries, different food companies, and bread types; therefore, the ingredient used in the frozen dough vary according to product characteristics and their requirement [12]. In the frozen dough making, the process typically includes ingredients mixing, molding, dividing, fermentation, packaging, and freezing [13, 14]. Based on the fermentation differences, frozen dough can be divided into two types: unfermented frozen dough and prefermented frozen dough [5, 8]. Unfermented frozen dough is produced by freezing the kneaded and shaped dough directly without a fermentation process, while prefermented frozen dough is fermented the kneaded and shaped dough to varying from 20% to 100% before frozen [15].

Therefore, in this chapter, we outline a clear procedure for unfermented frozen dough making using a Z-blade mixer, aiming to facilitate the reader understanding of the production steps of frozen dough preparation.

2 Materials

2.1 Dough Ingredients

An example of a dough formulation is provided in Table 1.

1. Wheat flour.
2. Water.

Table 1
An example of loaf bread dough formulation (adapt from [16])

Ingredient	Weight (g)
Wheat flour	1000
Water	600
Salt	15
Baking fat	7
Yeast	12
Ascorbic acid	0.2

3. Yeast.
4. Baking fat.
5. Salt.
6. Ascorbic acid (food grade).

2.2 Apparatus Required

1. Balance.
2. Containers.
3. Spatula.
4. Z-blade mixer (Morton Mixers, UK).
5. Polypropylene bags.
6. Freezer (−18°C).
7. Dough scraper.
8. Heat sealer.

3 Methods

All preparation should follow high hygienic standards, including handwashing before handing and preparation, personal cleanliness, keeping clean equipment and surface, and avoiding cross-contamination. The procedure of frozen dough is illustrated in Fig. 1.

1. Weigh 1000 g of wheat flour, 600 g of water, 15 g of salt, 12 g of yeast, 7 g of baking fat, and 0.2 g of ascorbic acid using an appropriate container.
2. Except for the water, pour all ingredients into the mixer tank and mix for 30 s at low speed (48 rpm, *see* **Note 1**).
3. Add water into the mixer tank and mix for 100 s at low speed (48 rpm) (*see* **Note 2**).
4. Mix again for 100 s at high speed (111 rpm).
5. Collect the bread dough from the mixer tank, separate into 460 g equal portions and shape the dough (*see* **Notes 3 and 4**).
6. Seal each portion of dough in a polypropylene bag and place it in a freezer (−18°C) for further analysis and/or use (*see* **Notes 5 and 6**).

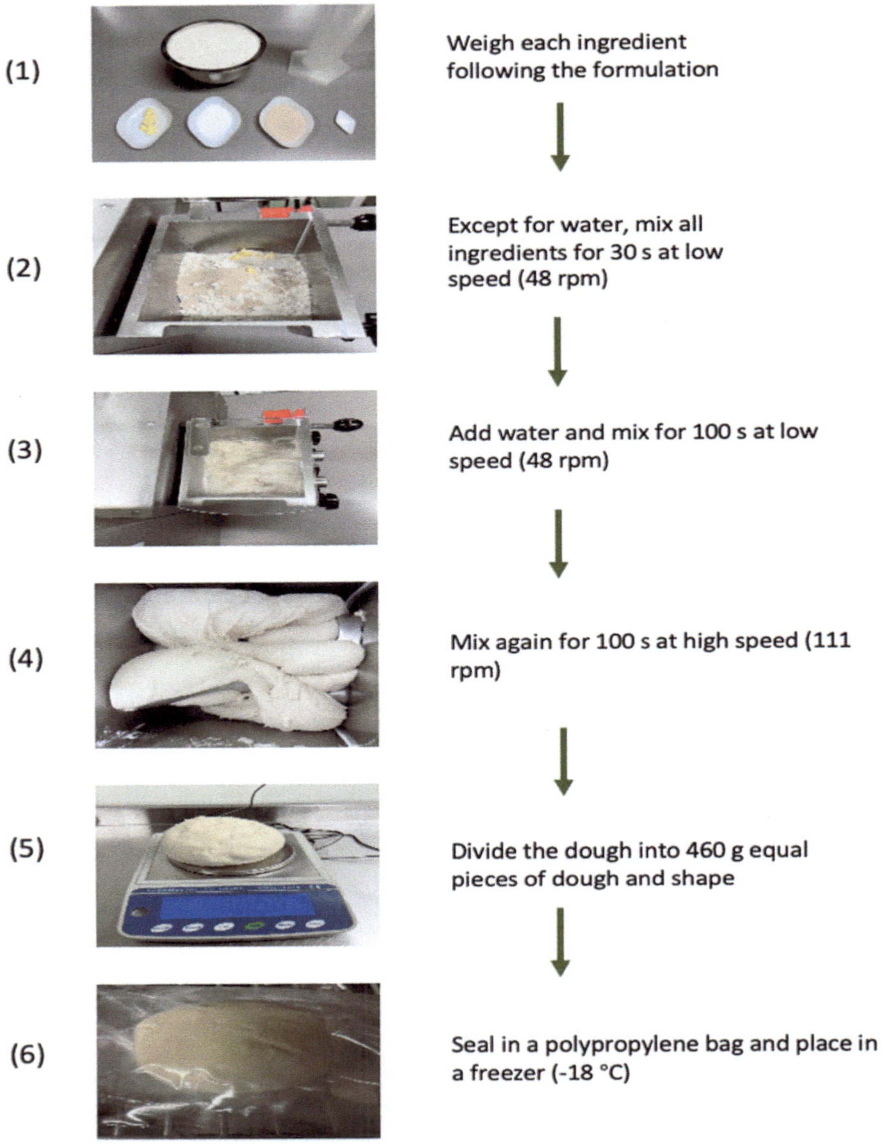

Fig. 1 Schematic and simplified representation of the steps of frozen dough preparation

4 Notes

1. Funnel could be used as a direct guide for the four during pouring into the small tank to reduce spillage.
2. After mixing, scrape the sides of the mixing tank with a spatula to ensure all ingredients are incorporated and returned to the dough.
3. The dough may stick, rub a small amount of oil on hands.

4. It is necessary to divide and shape dough relatively quickly to minimize the dough fermentation before freezing.
5. Frozen dough should be sealed as soon as possible to prevent drying out and minimize the dough fermentation.
6. When sealing, remove as much as air as possible to maintain dough quality, and ensure a complete seal to prevent air from entering.

References

1. Dong Y, Karboune S (2021) A review of bread qualities and current strategies for bread bioprotection: flavor, sensory, rheological, and textural attributes. Compr Rev Food Sci Food Saf 20:1937–1981. https://doi.org/10.1111/1541-4337.12717
2. Research and Markets, Bread Market—Global industry size, share, trends, opportunity, and forecast, 2020-2030F, (2025). https://www.researchandmarkets.com/reports/6169491/bread-market-global-industry-size-share?srsltid=AfmBOoqWoNoPDzGC9MsvgQzyjnPspwZgp26AcyQkE0crRU4mjzo--Yu1
3. Wang Y, Liu F, Xin M, Zhao F, Cheng Q, Guo J (2025) Comparative study of texture, gelatinization, retrogradation, structure, and digestibility in starch-hydrocolloid composites: effects of frozen and unfrozen treatments. Innov Food Sci Emerg Technol 105:104201. https://doi.org/10.1016/j.ifset.2025.104201
4. Wang L, Li Y, Guo Z, Wang H, Wang A, Li Z, Chen Y, Qiu J (2023) Effect of buckwheat hull particle-size on bread staling quality. Food Chem 405:134851. https://doi.org/10.1016/j.foodchem.2022.134851
5. Omedi JO, Huang W, Zhang B, Li Z, Zheng J (2019) Advances in present-day frozen dough technology and its improver and novel biotech ingredients development trends—a review. Cereal Chem 96:34–56. https://doi.org/10.1002/cche.10122
6. Li H, Lv Y, Zhang Y, Wang X, Li Z, Qu J (2024) Improvement of the freezing resistance characteristics of yeast in dough starter. Food Chem 458:140258. https://doi.org/10.1016/j.foodchem.2024.140258
7. Yang Z, Xu D, Zhou H, Wu F, Xu X (2022) Rheological, microstructure and mixing behaviors of frozen dough reconstituted by wheat starch and gluten. Int J Biol Macromol 212:517–526. https://doi.org/10.1016/j.ijbiomac.2022.05.144
8. Decock P, Cappelle S (2005) Bread technology and sourdough technology. Trends Food Sci Technol 16:113–120. https://doi.org/10.1016/j.tifs.2004.04.012
9. Wang Z, Tian X, Lin S, An Y, Zhao L, Li M (2023) Effects of commercial yeast on the quality of frozen dough and steamed bread in different regions of China. J Food Process Preserv 2023:1–10. https://doi.org/10.1155/2023/8873037
10. Avramenko NA, Tyler RT, Scanlon MG, Hucl P, Nickerson MT (2018) The chemistry of bread making: The role of salt to ensure optimal functionality of its constituents. Food Rev Int 34:204–225. https://doi.org/10.1080/87559129.2016.1261296
11. Zhang G, Wang Y, Liang Z, Guo J, Wu B (2025) Nutritional value and advances of mealworm (*Tenebrio molitor*) as a sustainable functional ingredient in bakery products: a key step toward sustainable diets. Future Foods 12:100716. https://doi.org/10.1016/j.fufo.2025.100716
12. Giannou V, Kessoglou V, Tzia C (2003) Quality and safety characteristics of bread made from frozen dough. Trends Food Sci Technol 14:99–108. https://doi.org/10.1016/S0924-2244(02)00278-9
13. Rosell CM, Gómez M (2007) Frozen dough and partially baked bread: an update. Food Rev Int 23:303–319. https://doi.org/10.1080/87559120701418368
14. Cappelli A, Bettaccini L, Cini E (2020) The kneading process: a systematic review of the effects on dough rheology and resulting bread characteristics, including improvement strategies. Trends Food Sci Technol 104:91–101. https://doi.org/10.1016/j.tifs.2020.08.008
15. Zhang M, Guo X, Sun X, Zhu K (2024) Frozen dough steamed products: deterioration mechanism, processing technology, and improvement strategies, Compr Rev Food Sci Food Saf 23:e70028. https://doi.org/10.1111/1541-4337.70028
16. Zhang G, Li Z (2024) Impact of melon seed oil cake with different particle sizes on bread quality. Food Prod Process Nutr 6:44. https://doi.org/10.1186/s43014-024-00225-9

Chapter 2

Determination of Rheological Properties

Run-Yang Zhang, Ling-Yu Qu and Hua-Min Liu

Abstract

Frozen dough technology is a promising approach that improves production efficiency and extends the shelf life of flour-based products. However, the processing of frozen dough, such as freezing, storage, and thawing, can affect its various components, including starch, gluten proteins, yeast, and lipids, ultimately impacting dough texture and baking quality. Therefore, systematic evaluation of the rheological properties of frozen dough is essential. This chapter details the materials and method operating procedures for measuring key parameters of rheological properties. These methods help elucidate the effects of freezing on dough structure and provide a scientific foundation for enhancing the quality of frozen dough.

Keywords Rheological Properties, Strength, Extensibility, Relaxation behavior

1 Introduction

The rheological properties of dough are directly related to its physical behavior during mixing, fermentation, shaping, and baking, ultimately influencing the texture and quality of baked products, and thus, serve as a key indicator for evaluating dough quality. Frozen dough technology notably enhances production efficiency, enabling the on-demand supply of fresh baked goods and meeting consumers' demand for freshness.

However, the frozen dough undergoes a series of complex physical and biochemical changes during freezing, storage, and thawing [1, 2]. These changes adversely affect the rheological properties of the dough. The primary effects include mechanical damage to the gluten network due to ice crystal formation and growth, water migration and redistribution, reduced yeast activity, and structural modifications in starch and proteins [3–5]. As a result, the dough might exhibit decreased viscoelasticity, weakened strength, and diminished gas retention capacity, ultimately leading to a negative impact on the volume, texture, and sensory quality of the final baked products [3–5]. For example, the rheological properties of wheat dough are significantly affected after freezing, with changes observed in both

Guoqiang Zhang and Jinying Guo (eds.), *Frozen Dough and Bread Quality*, Methods and Protocols in Food Science, https://doi.org/10.1007/978-1-0716-5206-0_2,

viscosity and elastic modulus [6]. The freezing process may damage the gluten network and reduce the dough's extensibility, leading to decreased water-holding capacity in bread made from the dough as frozen storage time increases and freezing rate decreases, along with declines in color, flavor, and overall acceptability [7, 8].

Therefore, measuring the rheological properties of frozen dough, e.g., strength, elasticity, viscosity, extensibility, and stress relaxation behavior, is crucial for evaluating product quality and baking performance. It not only helps reveal structural changes within the dough during the freeze-thaw process, but also guides the optimization of additives to maintain or improve dough quality.

2 Materials

2.1 Materials for Determination of Strength and Extensibility—Extensograph Test

1. Distilled or demineralized water.
2. Sodium chloride, analytical grade.
3. Optional release material: Rice flour or starch (*see* **Note 1**).
4. Spatula, thin, nonmetallic.
5. Conical flask, 250 mL.
6. Balance, accurate to 0.1 g.
7. Ultralow-temperature refrigerator with a freezing capacity below −40°C.
8. Refrigerator with a freezing range of −24°C to −16°C.
9. Electronic farinograph with following specifications:
   (a) Slow blade speed: (63 ± 2) r/min;
   (b) Speed ratio (fast blade: slow blade): (1.50 ± 0.01): 1;
   (c) The torque per FU (farinograph unit) of 300 g mixer is (9.8 ± 0.2) mN·m/FU [(100 ± 2) gf·cm/FU];
   (d) Water dosing system: For the 300 g mixer, a burette graduated from 135 mL to 225 mL in 0.2 mL divisions; The drainage time for both 0 mL to 225 mL does not exceed 20 s;
   (e) Thermostat with circulating water: (30 ± 0.2)°C.
10. Electronic extensograph with following specifications:
   (a) Balling unit speed: (83 ± 3) (r/min);
   (b) Moulder speed: (15 ± 1) (r/min);
   (c) Hook speed: (1.45 ± 0.05) cm/s;
   (d) Data automatically recorded by electronic system;
   (e) Thermostat with circulating water: (30 ± 0.2)°C.

2.2 Materials for Determination of Tenacity, Extensibility, Deformation Energy, and Curve Configuration Ratio—Alveograph Test

1. Distilled or demineralized water.
2. Sodium chloride solution, 0.025 g/mL (*see* **Note 2**).
3. Refined vegetable oil with low polyunsaturates (*see* **Note 3**).
4. Glass burette, of 50 mL in capacity, graduated in 1 mL divisions.
5. Balance, accurate to 0.01 g.
6. Ultralow-temperature refrigerator with a freezing capacity below −40°C.
7. Refrigerator with a freezing range of −24°C to −16°C.
8. Complete alveograph system: Composed of kneading machine, dedicated software, and alveograph. Technical specifications (*see* **Note 4**) are as follows:

   (a) Kneading blade speed: (60 ± 2) r/min;

   (b) Sheeting guide height: (12.0 ± 0.1) mm;

   (c) Sheeting roller, large diameter is (40.0 ± 0.1) mm, small diameter is (33.3 ± 0.1) mm;

   (d) Dough cutter inner diameter: (46.0 ± 0.5) mm;

   (e) Diameter of the aperture (moving plate opens): (55.0 ± 0.1) mm;

   (f) Distance between the fixed and moving plates after clamping: (2.67 ± 0.01) mm;

   (g) Volume of air automatically injected to detach the test piece prior to inflating the bubble: (18 ± 2) mL;

   (h) Inflation air flow rate: (96 ± 2) L/h.

2.3 Materials for Determination of Storage Modulus (G'), Loss Modulus (G"), Loss Tangent (tan δ), Creep Compliance (J), and Creep Recovery Compliance (J_r)—Rotational Rheometer Test

1. Distilled or demineralized water.
2. Sodium chloride, analytical grade.
3. Spatula, thin, nonmetallic.
4. Balance, accurate to 0.01 g.
5. Ultralow-temperature refrigerator with a freezing capacity below −40°C.
6. Refrigerator, with a freezing range of −24 °C to −16 °C.
7. Glass burette, 50 mL, 1 mL graduations.
8. Rotational rheometer: Composed of measuring device (viscometer, rheometer), measuring geometry (plate-plate geometry), and temperature control system.

3 Methods

3.1 Dough Strength and Extensibility—Extensograph Test

3.1.1 Sample Preparation

1. Wheat flour temperature: 25–30°C. Determine its moisture content as per ISO 712-1 [9].
2. Adjust the room temperature to between 18°C and 30°C. Maintain the thermostat and mixing bowl of farinograph at (30 ± 0.2)°C.
3. Install the 300 g mixing bowl on the farinograph's drive system. Enter test parameters and click "Start" to begin. Then, click "Start Test," and the instrument will automatically zero while the bowl runs empty. Opening the bowl cover activates the safety stop. During zeroing, the system sets the detected torque as the baseline (zero point) for all subsequent measurements.
4. Fill the farinograph burette using water maintained at (30 ± 0.5)°C.
5. In the farinograph software's test parameters window, enter the flour's moisture content. The software will calculate and display the required flour mass, equivalent to a 300 g sample (for a 300 g mixing bowl) at 14% moisture. Weigh the flour accurately to 0.1 g.
6. Add (6.0 ± 0.1) g of sodium chloride to a conical flask, and use a burette to add an amount of water equivalent to the water addition volume for the farinograph test reduced by 2–3% to dissolve it. For wheat flour with low water absorption, add less water.
7. Start the farinograph mixer and blend the flour at 63 r/min for 1 min. Add salt solution via funnel to bottom center hole of lid within 25 s—automatically recorded. Once dough forms, scrape bowl sides with a spatula to incorporate adhering particles, without stopping the mixer. Cover the bowl with the dedicated plastic lid. Consistency should be 500 ± 20 FU after (5 ± 0.1) min mixing. Stop mixing, turn off farinograph, and clean mixer.
8. Remove the dough from the mixing bowl, seal it in a polyethylene bag, and freeze it in an ultralow temperature freezer (below −40°C). When the core temperature reaches −18°C as detected by a thermocouple temperature sensor, transfer the dough to a standard freezer (with a freezing range between −24°C and −16°C) and store it at −18°C for 24 h. Afterwards, thaw the frozen dough on a plate at 25°C in the air. During the thawing process, use a thermocouple temperature sensor

to monitor the core temperature of the sample, and proceed to the next step when the temperature reaches 4°C.

9. Turn on the extensograph thermostat to maintain the cabinet at (30 ± 0.2)°C.
10. To ensure constant humidity, add water to the extensograph cradle support troughs until the bottom is fully covered. Allow supports, cradles, and clamps to equilibrate in the cabinet for at least 15 min before use.
11. Remove a support with cradles from extensograph cabinet and take off their clamps. Weigh a test piece of (150 ± 0.5) g, transfer it to the balling unit, and carry out 20 rotations of the plate. After removing the dough, guide it into the moulder centrally from the back, base first. Roll into the cradle center, secure with the clamp, then start a 45-min timer. Repeat the weighing, balling, moulding, and clamping steps for a second test piece. Place the cradle-support assembly with test pieces into the cabinet. For very sticky dough, lightly dust with rice flour or starch before moulding. If the dough shows significant elastic recovery that lifts the cradle's upper part, hold the clamps down for a few seconds to ensure proper fixation. Lastly, clean the farinograph mixer.

3.1.2 Determination

1. Exactly 45 min after clamping the first test piece, position the first cradle on the extensograph's balance arm with the bridge to the left to prevent contact with the stretching hook. Start the hook immediately, monitor the test piece, and remove the cradle once rupture occurs (*see* **Note 5**).
2. Remove the dough from the cradle and hook. Reball and remould the test piece, then reset the timer for 45 min.
3. Repeat the stretching procedure (Step 1) on the second test piece. Once the dough is removed from the cradle and hook, reapply the balling and moulding process (Sect. 3.1.1, Step 11).
4. Repeat the Steps 1–3, placing the formed test pieces back into the cabinet. Carry out this cycle beginning slightly after 90 min from the end of mixing.
5. Execute Step 1 again to stretch both test pieces sequentially. This final stretching is performed just over 135 min after mixing concludes.
6. To save time and conduct the measurement quickly, another operation procedure can be adopted. The rapid procedure differs from the standard procedure in resting time: stretching is performed at 30, 60, and 90 min after mixing, instead of 45, 90, and 135 min. Resulting curve properties and size differ from standard curves. This must be noted in the test report.

7. Results expression:
   (a) Energy: The area under recorded curve (Fig. 1), automatically computed by the instrument. Unit is square centimeters (cm^2), accurate to 1 cm^2.
   (b) Extensibility, *E*: The extensograph curve (Fig. 1) starts to rise until the maximum deformation occurs when the dough is stretched to the point of breaking. The unit for extensibility is millimeters (mm), accurate to 1 mm.
   (c) Resistance at constant deformation, R_{50}: Calculation begins from the moment the tensile resistance suddenly changes from zero when the tensile hook contacts the test specimen (Fig. 1). The height of the curve 5 cm after it starts to rise. The unit is EU.
   (d) Maximum resistance, R_m: Mean value of the maximum force levels recorded in the extensograph curves (Fig. 1) for both test pieces, acceptable only if the deviation between them does not exceed 15 % of the mean. Report in EU, accurate to 1 EU.
   (e) R_{50}/E or R_m/E: The ratio of resistance at constant deformation or the maximum resistance to the extensibility. They are accurate to 0.1.

3.2 Determination of Tenacity, Extensibility, Deformation Energy, and Curve Configuration Ratio—Alveograph Test

3.2.1 Sample Preparation

1. Ensure all apparatus components are clean. Maintain ambient temperature at 18–22°C and relative humidity at 50%–80%; Kneading machine: (24.0 ± 0.5)°C (start); Alveograph: (25.0 ± 0.5)°C (continuous); Wheat flour and sodium chloride solution: (20 ± 5)°C.

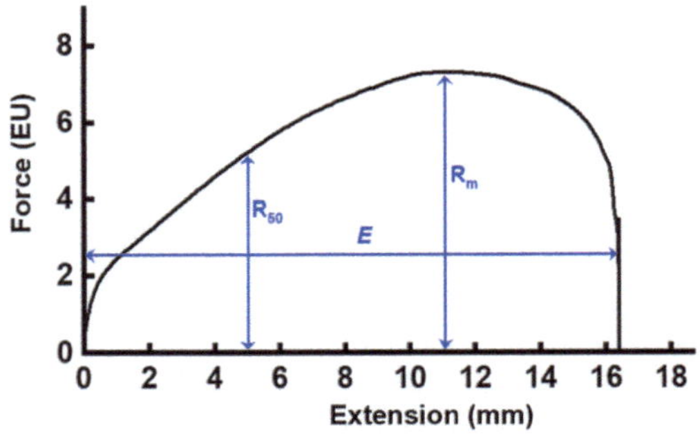

Fig. 1 The typical extensograph curve

2. Check that the Alveograph plate is horizontal; Check that the F-register is in place in the extrusion aperture to prevent any loss of wheat flour or salt solution leakage; Check the pneumatic circuit and flow settings by following the manufacturer's recommended procedure.
3. Measure flour moisture content as per ISO 712-1 [9].
4. Weigh (250 ± 0.50) g of wheat flour in the kneading machine, secure the lid with the locking device, switch on the motor. The sodium chloride solution is added by the fully automatic instrument (e.g., Alveolab model) or by the burette complete within 20–30 s (e.g., the AlveoNG and AlveoPC models). The addition amount is according to ISO 27971 [10], or calculate by the Eq. (1):

$$V_{NaCl} = 191.175 - \left(4.41175 \times H_f\right) \tag{1}$$

where H_f is the moisture content of the flour.

5. Stop kneading at 1 min, use a plastic spatula to mix the dry flour that has not been incorporated into the dough, which takes 1 min. Restart kneading at 2 min 0 s, continue kneading for 6 min, and stop at 8 min 0 s.
6. Freezing and thawing of dough: seal the dough in a polyethylene bag, and freeze it in an ultralow temperature freezer (below −40°C). When the core temperature reaches −18°C as detected by a thermocouple temperature sensor, transfer the dough to a standard freezer (with a freezing range between −24°C and −16°C) and store it at −18°C for 24 h. Afterwards, thaw the frozen dough on a plate at 25°C in the air. During the thawing process, use a thermocouple temperature sensor to monitor the core temperature of the sample, and proceed to the next step when the temperature reaches 4°C.
7. Rekneading: put the dough into the kneading machine and repeat Step 5.
8. Lift the gate of the dough extrusion outlet and tighten it, reverse the dough cutter, and drop little oil on the receiving plate at the extrusion outlet. Use a metal Spatula close to the dough extrusion outlet and quickly cutoff the first 10 mm slice of dough that comes out. Continue extruding the dough slices, use the spatula to lightly lift the end of each slice as it comes out to avoid sticking to the receiving plate. When reaching the marked position on the receiving plate, quickly cut with the spatula.
9. Test dough preparation: slide the dough slice on the receiving plate into the preoiled pressing slot (*see* **Note 6**). Roll the preoiled round presser continuously 12 times on the track of the pressing slot (back and forth six times). Use the preoiled round

cutter to cut in the center of the dough slice, remove the excess outer part, and move the round cutter with the small circular dough slice to the preoiled placement plate (*see* **Note 7**). Immediately place the samples in the 25°C blowing chamber in the order they are extruded, with the first one at the top and the rest placed downward in sequence.

3.2.2 Determination

1. Initial preparation: Perform the test 28 min after kneading begins. Check that the piston is in the raised position. Proceed in the order of extrusion of the test pieces.
2. First operation: placing the dough patty on the lower plate. For the AlveoNG and AlveoPC models:
   (a) First, unscrew the knurled ring and remove the pad. Unscrew the upper plate using the handle and reassemble it until it aligns with the three guide studs.
   (b) Turn the pad over and place it on the knurled ring, then oil both the fixed plate (being careful not to hit the bevel) and the inside of the pad.
   (c) Slide the dough piece onto the plate and center it by adjusting its edge. Replace the pad and secure it by screwing the knurled ring back in place.
   (d) Before conducting the test, lower the handle used for unscrewing the upper plate (first pull it upwards).
   (e) Flatten the dough piece by rotating the upper plate, which is equipped with the secured pad, for about 20 s or until it stops, ensuring no extra force is applied.
   (f) Finally, remove the knurled ring and the pad, and then lower the handle.

   For the Alveolab model: Take out the first dough piece and place it on the previously oiled (one to two drops) loading pad. Place a drop of oil on the dough piece. Press the start test button. The loading pad moves the dough piece up to the plate and blocks it at the test station.
3. Second operation: biaxial extension:
   For the AlveoNG and AlveoPC models: Press the [START/STOP TEST] key to initiate the Alveograph, which inflates the dough and measures its expansion. Immediately after the bubble bursts, press the [START/TEST STOP] key to halt the Alveograph and stop the airflow. Loosen the upper plate to fully release the tested dough piece. Repeat these steps, including placing and deforming the dough biaxially, for each of the remaining four dough pieces.
   For the Alveolab model: Once the dough is positioned and compressed, the plunger retracts automatically, and the device inflates the dough with air while recording bubble pressure. The resulting curve is displayed in real time. Test termination

occurs when the bubble bursts, detected instantly by the system. The plate then releases, and a squeegee removes the dough, depositing it into the collection tank. If the dough remains suspended, pressing the [dough stop] key allows for another pass of the squeegee. These operations should be repeated for each of the four remaining dough pieces.

4. Expression of results

   (a) Maximum pressure parameter, *P*: Relates to deformation resistance (tenacity). *P* value is the mean of the maximum ordinates of curve Fig. 2) multiplied by 1.1, and accurate to 0.1 units, in millimeters.

   (b) Mean abscissa at rupture, *L*: Relates to extensibility of dough. The value, in millimeters, is measure from the origin of the curves to the point corresponding vertically to the start of the pressure drop (Fig. 2).

   (c) Swelling index, *G*: This value is the square root of the volume of air, in milliliters, required to inflate the bubble until it ruptures. It is calculated using the following Eq. (2):

$$G = 2226\sqrt{L} \tag{2}$$

   (d) Elasticity Index, I_e: It is calculated using Eq. (3)

$$I_e = \left(\frac{P_{200}}{P}\right) \times 100 \tag{3}$$

   where P_{200} is the pressure inside the bubble when a volume of 200 mL of air has been injected into the test piece. P_{200} corresponds on the Alveogram to the height of the mean curve at L=40 mm, multiplied by the coefficient 1.1. I_e is accurate to 0.1%.

   (e) Curve configuration ratio, *P/L*: The ratio of P to L is the ratio of the curve shape. *P/L* is accurate to 0.01.

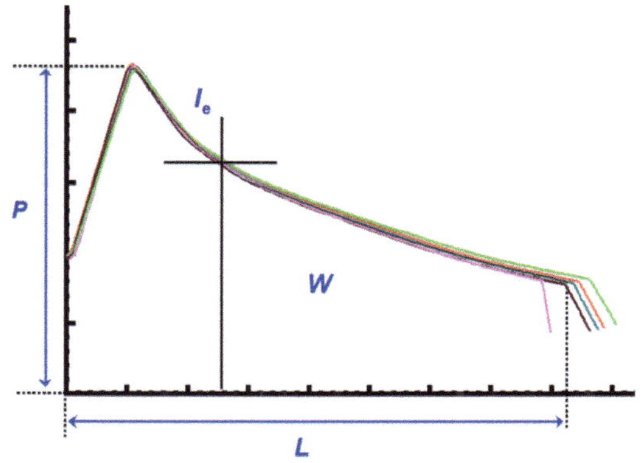

Fig. 2 The typical alveograph curve

(f) Deformation energy, W: Represent the baking strength of the flour and the work of deformation of 1 g of dough. W is calculated using the following Eq. (4):

$$W = 6.54 \times A \tag{4}$$

where A is the area under the mean curve. The coefficient 6.54 is valid for a constant air flow rate of 96 L/h. Express the W result to the nearest integer. It is expressed in units of 10^{-4} J.

3.3 Determination of Storage Modulus (G'), Loss Modulus (G"), Loss Tangent (tan δ), Creep Compliance (J), and Creep Recovery Compliance (Jr)—Rotational Rheometer Test

3.3.1 Sample Preparation

1. The dough is prepared according to the Steps 1–8 in Sect. 3.1.1. Then tested on a rheometer.

3.3.2 Determination

1. A sample (e.g., 8 g) is cut from the dough and avoid extra stress. It is positioned on a 40 mm parallel-plate fixture with a 2 mm gap, and excess dough is carefully trimmed (*see* **Note 8**). Temperature is controlled using a Peltier system, set to a specific value (e.g., 25°C or 30°C). The sample is allowed to rest for a period of time (e.g., 3, 5, and 10 min) in order to allow a builtup residual stress due to the clamp of the sample [11].
2. Dynamic strain sweep: To determine the Linear Viscoelastic Region (LVR or LVE), perform a strain-controlled oscillatory test at a constant frequency (e.g., 1 Hz and 10 Hz). Start from a very small strain amplitude (e.g., 0.01%) and gradually increase it (e.g., up to 100%). Monitor the storage modulus (G') and loss modulus (G") as a function of strain amplitude [8, 12]. LVR is the strain range where both G' and G" remain constant (form a plateau), independent of the applied strain. For subsequent tests on dough, select a constant strain amplitude within the LVR, typically well below the critical strain where G' begins to decrease significantly. A common practice is to choose a strain value at approximately 10–20% of the critical strain, or simply use a standard low strain (e.g., 0.5–2.5%) known to be within the LVR for similar dough systems (*see* **Note 9**).
3. Frequency sweep: Select a constant strain (e.g., 1.5%) within the LVR and set the frequency sweep range (e.g., 0.1–100 Hz).

Data-analysis software is used to determine the G', G", and tan δ. The unit of G' and G" is Pa.

4. Temperature sweep: Dough is heated over a temperature range (e.g., from 25 to 95°C) at a controlled rate (e.g., 2°C/min) and subjected to measurement at a frequency of 1 Hz and a strain (e.g., 1.5%) within the linear viscoelastic region (LVR) [8, 12]. The temperature-dependent variations in the sample's G' and G" are monitored and recorded.
5. Creep-recovery test: The dough is subjected to a constant shear stress (usually 50–80% of the LVR limit) at a certain temperature (e.g., 25°C) for 180 s. Then, release the stress (0 Pa) and monitor dough relaxation for 360 s. The time-dependent compliance (J) is recorded, from which the maximum creep compliance (J_{max}) at the end of the 180 s creep phase and the creep recovery compliance (J_r) at the end of the 360 s recovery phase are determined. The unit is Pa^{-1}.

4 Notes

1. Rice flour or starch is applied to prevent the dough from adhering to the moulder and rollers.
2. The sodium chloride solution is prepared by dissolving (25 ± 0.2) g of NaCl in distilled water and diluting to a final volume of 1000 mL. This solution should be used within 15 days and must be at a temperature of (20 ± 2)°C when applied.
3. Refined vegetable oil with low polyunsaturates, e.g., peanut oil, or olive oil (acid value ≤0.4 mg/g). It stores in a dark place and use within 3 months. Alternatively, liquid paraffin (acid value ≤0.05 mg/g; maximum 60 mPa·s at 20°C).
4. For instructions on operating the various registration systems, consult the manufacturer's manual.
5. In newer extensograph models, the hook automatically resets to its upper position. In older models, a switch is used to halt the hook after the sample breaks and to trigger its return to the starting position.
6. Pay attention to the direction of the dough slices when placing them; the extrusion direction should be consistent with the pressing slot direction.
7. The method is to tap the wrist on the table to make the dough slice fall, do not touch the sample. If the sample sticks to the pressing slot, slowly pry it up with the spatula and slide it onto the placement plate.

8. Sandpaper can be added to the lower plate to improve sample grip and prevent slippage. Mineral oil (paraffin) is applied to the sample edges to minimize moisture loss during testing.
9. Guarantee the test is performed within the linear viscoelastic region (LVR) of all samples.

References

1. Zhang H, Fan H, Xu X et al (2024) Deterioration mechanisms and quality improvement methods in frozen dough: an updated review. Trends Food Sci Tech 143:104251
2. Huang X, Wang S, Zhang M et al (2024) Enhancing frozen dough quality: investigating the impact of soy hull polysaccharide (SHP) on rheological properties and microstructure. Food Chem 454:139853
3. Zhou B, Dai Y, Guo D et al (2022) Effect of desalted egg white and gelatin mixture system on frozen dough. Food Hydrocoll132:107889
4. Guo L, He X, Wang Y et al (2024) Contribution of yeast freezing to the quality of frozen dough: rheological properties, protein depolymerization, gluten conformation and water state. Food Biosci 61:104866
5. Leray G, Oliete B, Mezaize S et al (2010) Effects of freezing and frozen storage conditions on the rheological properties of different formulations of non-yeasted wheat and gluten-free bread dough. J Food Eng 100:70–76
6. Yang Z, Xu D, Zhou H et al (2022) Rheological, microstructure and mixing behaviors of frozen dough reconstituted by wheat starch and gluten. Int J Biol Macromol 212:517–526
7. Obadi M, Xu B (2024) Characteristics and applications of plant-derived antifreeze proteins in frozen dough: a review. Int J Biol Macromol 255:128202
8. Gao K, Liu YX, Tan B et al (2022) An insight into the rheology and texture assessment: the influence of sprouting treatment on the whole wheat flour. Food Hydrocoll 125:107248
9. ISO (2024) Cereals and cereal products — Determination of moisture content — Part 1: Reference method. International Organization for Standardization, Geneva, Switzerland, pp 712-1
10. ISO (2023) Cereals and cereal products — Common wheat (*Triticum aestivum* L.) —Determination of Alveograph properties of dough at constant hydration from commercial or test flours and test milling methodology. International Organization for Standardization, Geneva, Switzerland, pp 27971
11. Jekle M, Becker T (2011) Dough microstructure: novel analysis by quantification using confocal laser scanning microscopy. Food Res Int 44:984–991
12. Hernández-Estrada ZJ, Rayas-Duarte P, Figueroa JDC et al (2014) Creep recovery tests to measure the effects of wheat glutenins on doughs and the relationships to rheological and breadmaking properties. J Food Eng 143:62–68

Chapter 3

Texture Analysis of Frozen Dough

Xi-Chuang Cheng, Qi-Wei Zhang and Hua-Min Liu

Abstract

The texture of frozen dough, which comprises multiple indicator attributes, serves as a key parameter for characterizing its internal mechanical and structure characteristics during processing. It is essential to examine the textural properties of frozen dough, as such analysis serves as a critical link between laboratory research, industrial production, and consumer perception. This chapter outlines a standardized operating procedure for texture analysis to ensure the accuracy, consistency, and comparability of test results.

Keywords Frozen dough, Texture profile analysis (TPA), Standardized protocol

1 Introduction

Analyzing the textural characteristics of frozen dough is crucial because it serves as the pivotal link between laboratory research, industrial production, and the consumer experience [1]. The processes of freezing, storage, and thawing pose a severe challenge to frozen dough, significantly impacting its internal structure. Relying solely on visual inspection or subjective experience makes it difficult to accurately evaluate these internal changes [2, 3]. Instrumental analysis is a highly significant and quantifiable analytical approach, primarily conducted using texture analyzers.

The texture profile analysis (TPA) mode of texture analyzers is a classical, objective, quantitative, and repeatable scientific method that converts microscopic damage to the gluten network into macroscopic measurements of hardness and elasticity. Furthermore, it enables the prediction of baking outcomes based on the condition of the raw dough [4, 5]. The principle of TPA involves applying a controlled, repeatable mechanical force to the sample to simulate the chewing action of the human mouth or the tactile sensation of the hand. At the same time, the response is precisely recorded, which reveals the relationship between force, deformation, and time. This process quantifies the macroscopic mechanical properties of the

Guoqiang Zhang and Jinying Guo (eds.), *Frozen Dough and Bread Quality*, Methods and Protocols in Food Science,
https://doi.org/10.1007/978-1-0716-5206-0_3,

dough and reflects changes in its internal microstructure [6, 7]. The following parameters of frozen dough can be obtained through TPA: hardness, springiness, cohesiveness, adhesiveness, chewiness, and resilience [8, 9].

The analysis of textural characteristics in frozen dough represents an indispensable core technology in the modern food industry, spanning from research and development to production [10]. Unfortunately, research on TPA conditions for frozen dough is limited, and there are no standardized TPA testing protocols in the baking industry. This chapter offers a detailed introduction to applying the TPA method to frozen dough.

2 Materials

2.1 Equipment and Supplies for Texture Profile Analysis

1. Analytical balance, accurate to ±0.0001 g.
2. Surgical knife.
3. Vernier caliper.
4. Equipment: texture analyzer (TA.XT Plus/PlusC, etc.) (*see* **Note 1**).

3 Methods

3.1 Texture Profile Analysis

3.1.1 Sample Preparation

Remove the frozen dough from the freezer and thaw it at 4°C for 12–16 h. Alternatively, thaw it at room temperature (25°C) until the core temperature reaches room temperature, which takes approximately 1–2 h. Preliminary testing is required to determine this time. Record the thawing conditions. Shape the thawed dough into uniform cylinders (e.g., Φ25×H25 mm) by hand or using molds. Avoid overkneading, as this can compromise the internal structure (*see* **Note 2**).

3.1.2 Analysis Procedure

1. Instrument calibration: Zero and calibrate the force and height settings of the texture analyzer prior to testing.
2. Test mode: Texture profile analysis (TPA).
3. Test speed: The pretest and test speeds are both 1.0 mm/s, and the posttest speed is 3.0 mm/s.
4. The compression level is 50% for soft samples and 40% for harder samples (*see* **Note 3**).
5. Trigger force: 5.00 g (*see* **Note 4**).

6. Testing process: Precisely position the sample at the center of the stage (*see* **Note 5**). Each sample must be tested at least six times. Maintain a stable ambient temperature (e.g., 25°C) throughout the testing process.

3.1.3 Data Collection and Analysis

1. **Data Acquisition:** The software automatically calculates and exports all TPA parameters, including hardness, springiness, cohesiveness, adhesiveness, chewiness, and resilience.

   Hardness is defined as the maximum force exerted during the initial compression and serves as a key parameter for evaluating bread staling. Increased hardness significantly affects texture.

   Springiness is the ratio of a deformed sample's height or volume to its original height or volume after the deformation force is removed. It is calculated by dividing the measured height of the second flush by the measured height of the first flush (Length 2/Length 1).

   Cohesiveness is defined as the internal cohesion within a sample, which ensures the maintenance of structural integrity throughout the sample. It is calculated as the ratio of the work areas required during the second and first impact processes (Area 2/Area 1).

   Adhesiveness refers to the force required for a probe to overcome the attractive forces between two surfaces upon making contact with a sample. It can be quantified by measuring the ratio of the negative peak area to the penetration area during the initial thru-down stage (Area 3/Area 4).

   Chewiness refers to the amount of energy required to masticate a semisolid sample into a state suitable for swallowing. It can be calculated using the formula: hardness × cohesiveness × springiness).

   Resilience indicates the extent to which a deformed specimen returns to its original state under the same conditions. It is calculated by dividing the area during the "recovery" stage of the first stamping process by the area during the downward stamping stage (Area 5/Area 4).

2. **Data Processing**

Exclude the obvious outliers. Calculate the mean and standard deviation (SD) or standard error of the mean (SEM) for each TPA parameter.

3.1.4 Statistical Analysis

Use statistical software (e.g., SPSS, SAS, or Origin) to perform an analysis of variance (ANOVA) and significance tests (e.g., Duncan's or Tukey's test) to determine if differences between different samples are statistically significant (typically, at a significance level of $p<0.05$).

4 Notes

1. Force sensors: The texture analyzer is equipped with that are rated at 5, 25, or 50 kg, depending on the hardness of the sample. Probe: Cylindrical probes (P/36 R or P/50) are recommended, with a 36 mm diameter suitable for most dough samples.
2. The primary requirement for testing is maintaining sample consistency. Samples must be uniform in size, shape, and weight.
3. This parameter significantly impacts the results and must be set correctly. 50% is a common setting, but adjustments may be needed for extremely soft or hard products, such as 30% or 70%.
4. Ensure consistent starting points for each test.
5. Make sure that the sample center is aligned with the probe and that the surface is level.

References

1. Yadav DN, Patki PE, Sharma GK et al (2009) Role of ingredients and processing variables on the quality retention in frozen bread doughs: a review. J Food Sci Technol 46(1): 12–20
2. Dou XL, Lv MS, Ren XY et al (2023) Test conditions of texture profile analysis for frozen dough. Ital J Food Sci 35(4):58–68
3. Wei Q, Zhang G, Xie J (2025) Modulatory effects of carboxymethyl chitosan on the textural and quality attributes of frozen rice dough and corresponding rice bread. Food Hydrocoll 162:111021
4. Liu HY, Xing YA, Liu DZ et al (2023) Quality deterioration of frozen dough bread during terminal freezing and thawing: from the insight of moisture and starch properties. Food Biosci 56:103435
5. Bekiroglu H, Ozulku G, Sagdic O (2023) Effects of casein hydrolysate prepared with savinase on the quality of bread made by frozen dough. Foods 12(20):3845
6. Shin SH, Choi WS (2021) Variation in significant difference of sausage textural parameters measured by texture profile analysis (TPA) under changing measurement conditions. Food Sci Anim Resour 41(4):739–747
7. Cakir E (2011) Understanding textural properties of protein-based soft-solid structures using oral processing. Doctoral dissertation, North Carolina State University
8. Luo WH, Sun DW, Zhu ZW et al (2018) Improving freeze tolerance of yeast and dough properties for enhancing frozen dough quality—a review of effective methods. Trends Food Sci Technol 72:25–33
9. Kondakci T, Zhang JW, Zhou WB (2015) Impact of flour protein content and freezing conditions on the quality of frozen dough and corresponding steamed bread. Food Bioprocess Technol 8(9):1877–1889
10. Arias AC, Bobadilla CAF, Domínguez CMZ (2024) Cryoprotectants for frozen dough: a review. Food Biophysics 19(1):18–28

Chapter 4

Dough Pasting Analysis

Jinying Guo and Yange Tian

Abstract

Dough pasting properties are critical determinants in the quality of flour-based products, especially during thermal processing such as baking, steaming, or extrusion. These properties reflect the structural changes and viscosity evolution of dough as it undergoes heating and cooling, primarily driven by starch gelatinization and retrogra-dation. Understanding the pasting behavior of dough enables the optimization of processing parameters and product formulations. This chapter introduces a standard method for characterizing the pasting properties of dough using a rapid visco-analyzer (RVA), which simulates the thermal and shear conditions encountered during actual processing. The test records the variation in viscosity under a controlled temperature profile, providing key parameters such as peak viscosity, trough viscosity, breakdown, final viscosity, and pasting temperature.

Keywords Dough pasting properties, Rapid visco-analyzer (RVA), Starch gelatinization, Viscosity, Pasting temperature

1 Introduction

The pasting properties of dough are one of the core factors determining the final quality of various flour-based foods. During thermal processing stages such as baking, steaming, or extrusion, understanding and controlling the pasting behavior of dough is crucial for optimizing product texture, structure, stability, and even sensory attributes. Starch, as the primary carbohydrate in dough, undergoes gelatinization and retrogradation processes that dominate the evolution of its pasting properties. When heated in the presence of water, starch granules swell, absorb water, and leach amylose, leading to a rapid increase in viscosity—a process known as gelatinization. The temperature at which viscosity begins to increase marks the initiation of starch gelatinization [1]. Further heating or shearing may cause the disintegration of starch granules and a decrease in viscosity (breakdown). Upon cooling, the reassociation of starch molecules (retrogradation) results in an increase in final viscosity. These changes directly influence the texture, water absorption, stability, and overall

Guoqiang Zhang and Jinying Guo (eds.), *Frozen Dough and Bread Quality*, Methods and Protocols in Food Science,
https://doi.org/10.1007/978-1-0716-5206-0_4,

quality of the final product [2, 3]. Therefore, precise characterization of the pasting properties of dough is invaluable for predicting final product quality, guiding formulation adjustments, and optimizing processing parameters.

Among various analytical techniques, the Rapid Visco-Analyzer (RVA) has become an authoritative tool for characterizing the pasting behavior of starch and dough. The RVA simulates typical heating-holding-cooling processes encountered in food processing within a short test cycle by precisely controlling the temperature profile and applying constant shear rate. It offers outstanding advantages such as small sample requirement, rapid testing, high reproducibility, and ease of operation [4]. The test records the variation in viscosity under a controlled temperature profile, providing key parameters such as peak viscosity, trough viscosity, breakdown, final viscosity, and pasting temperature [5]. The resulting pasting curve provides valuable insights into the thermal stability of the dough, the swelling capacity of the starch, and the tendency toward retrogradation. It has become a standard tool for screening flour quality, evaluating the effects of additives, and predicting product performance in industrial applications.

This chapter provides a detailed and reproducible experimental protocol for assessing the pasting properties of dough using the RVA, including sample preparation, instrument operation, and data interpretation.

2 Materials

2.1 Reagents

1. Wheat flour
2. Distilled water
3. Sodium chloride solution (optional, for standardized testing conditions)

2.2 Apparatus

1. Ensure all instruments are clean and dry before use
2. Rapid visco-analyzer (RVA) (RVA4800, Perten Instruments, Newport Scientific)
3. RVA test canisters and stirrers
4. Analytical balance (accuracy: 0.01 g)
5. Measuring cylinder
6. Glass rod
7. Beaker

3 Methods

Follow laboratory safety regulations during operation.

3.1 Sample Preparation

1. Weigh the flour mixture or freeze-dried dough powder and place it into a dedicated RVA aluminum canister [6] (recommended 3–5 g) (*see* **Note 1**).
2. Precisely add 25 mL of distilled water to the canister (usually total weight of the sample and is 28–30 g).
3. Stir the mixture homogenously with a glass rod to avoid lump formation (*see* **Note 2**).

3.2 RVA Testing Procedure

1. Select the standard pasting profile [7] (e.g., AACC Method 76-21).
2. Start temperature: 50°C.
3. Hold at 50°C for 1 min.
4. Heat from 50°C to 95°C at a rate of 12°C/min.
5. Hold at 95°C for 2.5 min.
6. Cool from 95°C to 50°C at a rate of 12°C/min.
7. Hold at 50°C for 2 min.
8. Set the stirring speed at 160 rpm during the entire test.
9. Start the test and allow the RVA to record viscosity (in cP or Pa·s) continuously throughout the temperature program.
10. After the test, remove the canister carefully (*see* **Note 3**).

3.3 Data Recording and Analysis

1. From the pasting curve, identify the following parameters (see Fig. 1 for reference):

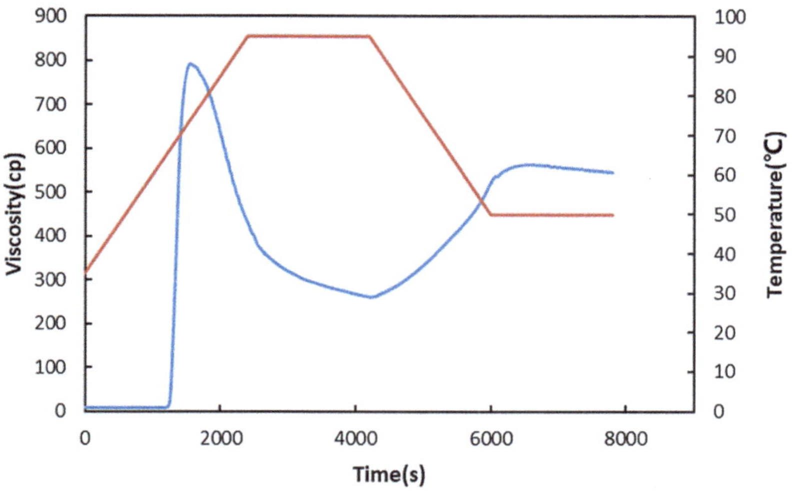

Fig. 1 Rapid viscosity analysis curves of dough with different fermentation methods

2. Pasting temperature: Temperature at which viscosity begins to increase significantly. This marks the initiation of the starch gelatinization process.
3. Peak viscosity: Maximum viscosity during heating. Reflects the water-binding capacity of the starch.
4. Trough viscosity: Minimum viscosity after the peak, usually at the end of the 95°C holding period.
5. Breakdown: Peak viscosity—trough viscosity. Reflects the extent of amylose leaching and the thermal stability of the starch [8].
6. Final viscosity: Viscosity at the end of the test (after cooling to 50°C), which is influenced by the degree of amylose gelatinization.
7. Setback: Final viscosity—trough viscosity. Reflects the extent of amylose retrogradation occurring during the cooling process [9].
8. Export the time-temperature-viscosity data in.xls or.csv format.
9. Use plotting software (e.g., Origin, Excel) to visualize the pasting curve.
10. Compare pasting parameters among different samples and perform statistical analysis using SPSS or similar software.

4 Notes

1. Sample weight and moisture content must be strictly controlled for reproducibility. Some protocols use ground dry flour + water; when using formed dough, adjust total solids to match standard test conditions.
2. Measure immediately after mixing.
3. Clean the RVA canister and stirrer thoroughly immediately after use to prevent starch residue buildup.

References

1. Zhang Q, Sun P, Fan S et al (2024) *Gelidium amansii* polysaccharide–based flour as a novel ingredient in gluten-free dough: effects on the rheological and thermomechanical properties. Food Bioprocess Technol 17:4040–4052. https://doi.org/10.1007/s11947-024-03370-8
2. Guo J, Liu F, Gan C, Wang Y, Wang P, Li X, Hao J (2022) Effects of konjac glucomannan with different viscosities on the rheological and microstructural properties of dough and the performance of steamed bread. Food Chem 368:130853. https://doi.org/10.1016/j.foodchem.2021.130853
3. Lu C, Guo J, Li P, Bai Z, Cui G, Li P (2025) Physicochemical properties and *in vitro* digestion of quinoa starch induced by combination of ultrasound and konjac glucomannan. Food Chem 463:141380. https://doi.org/10.1016/j.foodchem.2024.141380

4. Jinying G, Chengyan W, Changying L, Ping W (2020) Effect of konjac glucomannan on gelatinization, retrogradation, and gelling properties of frozen wheat starch. Starch-Stärke 73:1911386. https://doi.org/10.1002/star.202000025
5. Espinosa-Ramírez J, De la Rosa-Millan J, Pérez-Carrillo E, Serna-Saldívar SO (2021) Assessment of the quality of fresh nixtamalized maize doughs with different degrees of cooking and milling: a comparison of Mixolab and RVA analyses. J Cereal Sci 102:103321. https://doi.org/10.1016/j.jcs.2021.103321
6. Zeng F, Yang Y, Liu Q, Yang J, Jin Z, Jiao A (2023) Effect of fermentation methods on properties of dough and whole wheat bread. J Sci Food Agric 103:4876–4886. https://doi.org/10.1002/jsfa.12565
7. Yu Z, Xiaoxuan W, Xiao G (2022) Effects of adding quinoa flour on the composite wheat dough: a comprehensive analysis of the pasting, farinograph and rheological properties. Int J Food Sci Technol 57(11):7099–7106. https://doi.org/10.1111/ijfs.16049
8. Pandey S, Gujral HS (2024) Mixolab thermo-mechanical behaviour of millet flours and their cookie doughs, flour functionality and baking characteristics. J Cereal Sci 116:103866. https://doi.org/10.1016/j.jcs.2024.103866
9. Mohamed AA, Alamri MS, Hussain S, Ibraheem MA, Qasem AAA, Shamlan G, Ababtain IA (2022) Effect of ziziphus and cordia gums on dough properties and baking performance of cookies. Molecules 27:3066. https://doi.org/10.3390/molecules27103066

Chapter 5

Dough Thermal Analysis

Jinying Guo and Jialu Chen

Abstract

The thermal behavior of dough is a critical determinant of the final technical and sensory quality of flour-based products. During thermal processing, gluten protein denaturation and starch gelatinization occur as concurrent phenomena that collectively govern the development of the product's macroscopic structure and texture. This chapter details an experimental methodology using differential scanning calorimetry (DSC) as the primary tool for precise characterization of the associated physicochemical transitions. The systematic analysis presented herein aims to establish a fundamental structure–function understanding, thereby providing a theoretical foundation for quality control, process innovation, and precision manufacturing in the industry.

Keywords Dough, Thermal processing, Differential scanning calorimetry, Characterization

1 Introduction

In the processing of flour-based products, the thermal properties of dough exert a decisive influence on the final technical and sensory quality of the end product [1, 2]. During thermal processing, the denaturation of gluten proteins and the gelatinization of starch occur as two core, concurrent processes that collectively drive the formation of the dough's macroscopic structure and texture [3, 4]. Among these, the denaturation behavior of gluten is particularly critical for establishing a stable structural framework [5]. An in-depth investigation into the thermal characteristics of dough is central to understanding the transformation mechanism from raw materials to finished products. The intrinsic thermal properties of dough directly dictate its structural evolution pathway during heat treatments such as baking and steaming, thereby governing the final texture and morphology of the product. Consequently, a systematic analysis of dough's thermal properties not only aids in elucidating the scientific basis of the structure–function relationship during processing but also provides a crucial theoretical foundation and technical pathway for quality control, process innovation, and precision manufacturing

Guoqiang Zhang and Jinying Guo (eds.), *Frozen Dough and Bread Quality*, Methods and Protocols in Food Science,
https://doi.org/10.1007/978-1-0716-5206-0_5,

of flour-based products. Differential scanning calorimetry (DSC) serves as an effective tool for studying such thermal transition behaviors, enabling the precise characterization of the physicochemical changes that dough undergoes during heating [6]. Based on the aforementioned background, this chapter will detail the experimental procedures for determining the thermal properties of dough samples using differential scanning calorimetry (DSC).

2 Material

2.1 Reagents

1. Wheat flour
2. Distilled water
3. Yeast

2.2 Apparatus Required

Check that all instruments are clean and dry before use

1. Differential scanning calorimeter (DSC Q2000, Mettler-Toledo, Switzerland)
2. Aluminum Hermetic Pan
3. Analytical balance (accuracy: 0.01 g)
4. Mixer (e.g., Farinograph, Mixolab, or small-scale laboratory mixer)
5. Proofing chamber (for dough resting, optional)
6. Timer

3 Methods

All operations must comply with laboratory safety regulations, and waste materials must be disposed of appropriately.

3.1 Sample Digestion

1. The dough should be prepared as per the recipe (*see* **Note 3**).
2. Mix using a dough hook for the specified duration, until the gluten is fully developed.
3. The prepared dough was freeze-dried and passed through a 100-mesh sieve.
4. Approximately 3 mg of freeze-dried dough powder was weighed into an aluminum crucible (*see* **Notes 1, 4, and 6**). Distilled water was then added at a 1:3 (w/w) powder-to-water ratio [7].
5. The prepared sample was subsequently equilibrated at 4°C for 12 h [8].

3.2 Differential Scanning Calorimetry (DSC) Analysis

1. The aluminum crucible containing the sample was placed into the differential scanning calorimeter using tweezers.
2. The sample crucible is placed on the right, and the empty (reference) crucible on the left.
3. Purge with nitrogen gas to prevent oxidation (select atmosphere as required) (*see* **Notes 7 and 8**).
4. The heating temperature was from 30°C to 120°C at a rate of 10°C/min (*see* **Note 2, 5, and 6**).
5. Record the denaturation peak temperature (Tp) and enthalpy values (ΔH).

3.2.1 Data Processing

Export the heat flow curve (DSC curve) of the sample during heating, along with the data in.xls format.

4 Notes

1. The sample must be homogeneous, as clumping can lead to inaccurate measurements.
2. A blank experiment should be performed prior to measurement for baseline subtraction.
3. Each test was performed in at least triplicate to ensure the reliability and reproducibility of the results.
4. Ensure precise weighing of the sample mass.
5. Ensure the set temperature does not exceed the maximum operating temperature of the crucible (e.g., ≤600°C for aluminum crucibles).
6. Ensure the crucible material and sealing method are compatible with the test temperature and sample properties.
7. The required test gas (e.g., inert, oxidizing, or reactive) must be clearly defined and correctly configured.
8. Ensure that the gas line is properly sealed and the flow rate is stable.

References

1. Zhang G, Guo J, Zhao F, Bai Z, Zhang L (2024) Peptide incorporation in frozen dough and steamed bread: characteristics and structure-function relationship. LWT-Food Sci Technol 205:116551. https://doi.org/10.1016/j.lwt.2024.116551.2024.7.25
2. Lu C, Guo J, Li P, Bai Z, Cui G, Li P (2025) Physicochemical properties and in vitro digestion of quinoa starch induced by combination of ultrasound and konjac glucomannan. Food Chem 463:141380. https://doi.org/10.1016/j.foodchem.2024.141380.2024.9.20
3. Duan B, Guo J, Li P, Cheng B, Zhang G, Bai Z, Luo D, Cui C (2024) Moisture, rheology and structure of deacetylated konjac glucomannan-treated dough and the performance of steamed bread under frozen storage.

LWT-Food Sci Technol 197:115941. https://doi.org/10.1016/j.lwt.2024.115941.2024.3.4
4. Cheng B, Guo J, Duan B, Zhao F, Lu C, Bai Z, Wang L, Xing S (2024) A comparative study of molecular structure, rheology, and thermal characteristics of deacetylated konjac glucomannan-treated gluten during frozen storage. J Cereal Sci 117:103914. https://doi.org/10.1016/j.jcs.2024.103914.2024.7.17
5. Wang Y, Guo J, Wang C, Li Y, Bai Z, Luo D, Hu Y, Chen S (2023) Effects of konjac glucomannan and freezing on thermal properties, rheology, digestibility and microstructure of starch isolated from wheat dough. LWT-Food Sci Technol 177:114588. https://doi.org/10.1016/j.lwt.2023.114588
6. Guo J, Liu F, Gan C, Wang Y, Wang P, Li X, Hao J (2022) Effects of Konjac glucomannan with different viscosities on the rheological and microstructural properties of dough and the performance of steamed bread. Food Chem 368:130853. https://doi.org/10.1016/j.foodchem.2021.130853
7. Guo J, He Y, Liu J, Wu Y, Wang P, Luo D, Xiang J, Sun J (2021) Influence of konjac glucomannan on thermal and microscopic properties of frozen wheat gluten, glutenin and gliadin. Innov Food Sci Emerg Technol 74:102866. https://doi.org/10.1016/j.ifset.2021.102866
8. He Y, Guo J, Ren G, Cui G, Han S, Liu J (2020) Effects of konjac glucomannan on the water distribution of frozen dough and corresponding steamed bread quality. Food Chem 330:127243. https://doi.org/10.1016/j.foodchem.2020.127243

Part II

Bread Baking and Analysis

Check for updates

Chapter 6

Frozen Dough Bread Baking

Guoqiang Zhang

Abstract

Bread is one of the popular and important staple foods in the human diet. The development of frozen dough technology eliminates the initial dough preparation step in the traditional bread baking process, thereby reducing the production burdens and offering a convenient method for producing freshly bread on demand. However, it is well known that dough preparation alone is not enough to obtain the final bread product. In addition to frozen dough preparation, the subsequent processing steps play an important role in transforming raw dough into the final bread, imparting bread with the unique and attractive sensory properties and flavor. Therefore, this chapter provides baking procedures for frozen dough bread making.

Keywords Frozen dough, Frozen dough bread, Bread baking, Baking process, Baking products

1 Introduction

Bread is an important part of human diet and is one of the most consumed and popular staple foods in the world [1, 2]. The introduction of frozen dough in bread making procedure eliminates the initial dough preparation step and allows for prolonged preservation, thereby reducing the production burden and providing greater flexibility to produce freshly baked bread as needed; however, the frozen dough may need to be thawed before further processing, depending on the type of the final bread product [3, 4]. It has been reported that for frozen croissant dough, thawing is not required and can be transferred directly from the freezer to the oven [5]. On the other hand, during the thawing processing, the quality of frozen dough bread is associated with thawing method; therefore, selecting appropriate thawing method is essential to maintain the quality of frozen dough [6]. Ambient temperature thawing and refrigerator thawing are traditional and the most used thawing methods in frozen dough [7].

After thawing, fermentation (referring to proofing) the dough is one of the important steps in bread baking, which affects the quality of the final products, such as texture and flavor [8, 9]. During dough

Guoqiang Zhang and Jinying Guo (eds.), *Frozen Dough and Bread Quality*, Methods and Protocols in Food Science,
https://doi.org/10.1007/978-1-0716-5206-0_6,

fermentation, yeast, is regarded as leavening agent in baking products, consumes sugar to produce alcohol, carbon dioxide, and other metabolites (e.g., organic acids and aroma compounds), making dough expansion, and providing unique flavor and taste [8, 10]. In general, the fermentation temperature and relative humidity for dough range from 25°C to 45°C and 70%–90%, respectively, while the fermentation time depends on multiple factors, such as the type of product, dough preparation method, dough ingredients, and concentration of yeast [11, 12]. Notably, in the case of prefermented frozen dough, the fermentation step after thawing could be skipped, and the dough can be baked directly [13].

Baking is the last but most important step in bread making, converting the raw dough into finished baked products with attractive and desirable sensory characteristics (e.g., browning and crunchy crust, enticing flavor, and soft and resilient texture) [14, 15]. During baking, a complex series of physical, chemical, and biological changes in dough with temperature increasing, including water evaporation, volume expansion, protein denaturation, starch gelatinization, and Maillard reaction, whereas baking temperature, time, and ingredient used are considered key factors [16–18]. The bread baking temperature typically in a range between 160°C and 250°C with variations depending on the type of product, baking technology, and machine used [19]. In case of frozen dough bread products, the baking process is not significantly difference from that of traditional bread products.

In light of the above, this chapter provides a detailed baking procedures for frozen dough bread making, including the thawing methods for frozen dough and subsequent baking steps. The procedure is described in an example of unfermented frozen dough for loaf bread baking, guiding the reader through the conversion process from frozen dough to the final baked bread.

2 Materials

Unfermented frozen dough is used.

1. Loaf tin.
2. Oven gloves.
3. Spray oil.
4. Mono bread moulder (Mono Equipment, Swansea, UK).
5. Proofing oven (ARM/93 proof oven, Salva, Lezo, Spain) (*see* **Note 1**).
6. Deck oven (3STA 4676, Polin Stratos, Verona, Italy) (*see* **Note 1**).
7. Polypropylene bags.

3 Methods

This protocol uses unfermented frozen dough for loaf bread baking, therefore, thawing and fermentation steps are required.

3.1 Ambient Thawing

Remove the frozen dough from the freezer (−18°C) and place it in the container or on a baking tray, covered with film to avoid drying. Afterwards, thaw the frozen dough in the room temperature for 60 min.

3.2 Refrigerator Thawing

Remove the frozen dough from the freezer (−18°C) and place it in a fridge (4°C) for 12 h.

3.3 Baking Procedure [20]

The baking procedure is illustrated in Fig. 1.

1. Spray the insides of the loaf tin with a small amount of oil and then place the thawed dough in the loaf tin (*see* **Note 2**).
2. Place the loaf tin into the proofing oven for initial proofing at 40°C and 80% relative humidity for 20 min.
3. After the initial proofing, shape the dough using mono bread moulder.
4. Place the loaf tin back into the proofing oven for a second proofing for 20 min.
5. After second proofing, place the loaf tin into the oven deck for baking at 220°C for 20 min (*see* **Note 3**).

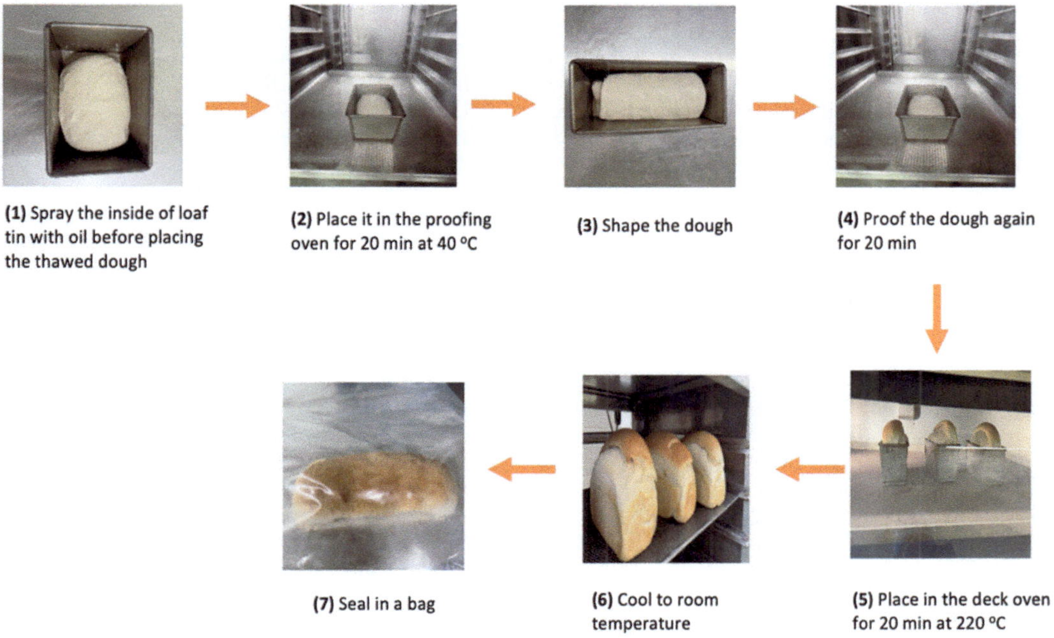

Fig. 1 Schematic and simplified representation of the baking procedure for frozen dough bread

6. After baking, take the baked bread from the baking deck and place it on a cooling rack to cool to room temperature (*see* **Note 3**).
7. Seal the bread in a polypropylene bag (*see* **Note 4**).

4 Notes

1. Preheat the proofing oven and deck oven before using.
2. Spray a light, even layer on the bottom and all sides of the loaf tin to prevent the dough and/or bread from sticking to the tin.
3. The deck oven and baked bread are very hot and can cause burns, always use oven gloves during operation.
4. Before sealing, bread should be cooled to room temperature to prevent condensation of moisture in the bag.

References

1. Dong Y, Karboune S (2021) A review of bread qualities and current strategies for bread bioprotection: flavor, sensory, rheological, and textural attributes. Compr Rev Food Sci Food Saf 20:1937–1981. https://doi.org/10.1111/1541-4337.12717
2. Zhang G, Li Z (2024) Impact of melon seed oil cake with different particle sizes on bread quality. Food Prod Process Nutr 6:44. https://doi.org/10.1186/s43014-024-00225-9
3. Wang M, Chen X, Yang Z, Wang M, Shang S, Fu B, Wen C (2025) Effect of acetylated distarch adipate on fermentation properties of freeze-thawed post-fermented frozen dough and bread characteristics. J Cereal Sci 121:104094. https://doi.org/10.1016/j.jcs.2024.104094
4. Zhao B, Hou L, Liu T, Liu X, Fu S, Li H (2023) Insight into curdlan alleviating quality deterioration of frozen dough during storage: fermentation properties, water state and gluten structure. Food Chem X 19:100832. https://doi.org/10.1016/j.fochx.2023.100832
5. Omedi JO, Huang W, Zhang B, Li Z, Zheng J (2019) Advances in present-day frozen dough technology and its improver and novel biotech ingredients development trends—a review. Cereal Chem 96:34–56. https://doi.org/10.1002/cche.10122
6. Zhang M, Guo X, Sun X, Zhu K (2024) Frozen dough steamed products: deterioration mechanism, processing technology, and improvement strategies. Compr Rev Food Sci Food Saf 23:e70028. https://doi.org/10.1111/1541-4337.70028
7. Zhang F, Guo J, Li P, Zhao F, Yang X, Cheng Q, Elmore JS, Wang P, Cui C (2024) Effects of different thawing methods on physical and physicochemical properties of frozen dough and quality of corresponding steamed bread. Food Chem 447:138932. https://doi.org/10.1016/j.foodchem.2024.138932
8. Struyf N, Van der Maelen E, Hemdane S, Verspreet J, Verstrepen KJ, Courtin CM (2017) Bread dough and Baker's yeast: an uplifting synergy. Compr Rev Food Sci Food Saf 16:850–867. https://doi.org/10.1111/1541-4337.12282
9. Obadi M, Xu B (2024) Characteristics and applications of plant-derived antifreeze proteins in frozen dough: a review. Int J Biol Macromol 255:128202. https://doi.org/10.1016/j.ijbiomac.2023.128202
10. Huang S, Miskelly D (2019) Steamed bread—a review of manufacturing, flour quality requirements, and quality evaluation. Cereal Chem 96:8–22. https://doi.org/10.1002/cche.10096
11. Suchintita Das R, Tiwari BK, Garcia-Vaquero M (2023) The fundamentals of bread making: the science of bread. Traditional European Breads. Springer International Publishing, Cham, pp 1–40. https://doi.org/10.1007/978-3-031-23352-4_1
12. Huang L, Wan J, Huang W, Rayas-Duarte P, Liu G (2011) Effects of glycerol on water properties and steaming performance of pre-fermented frozen dough. J Cereal Sci 53: 19–24. https://doi.org/10.1016/j.jcs.2010.07.006

13. Olakanmi SJ, Jayas DS, Paliwal J (2023) Applications of imaging systems for the assessment of quality characteristics of bread and other baked goods: a review. Compr Rev Food Sci Food Saf 22:1817–1838. https://doi.org/10.1111/1541-4337.13131
14. Purlis E (2012) Baking process design based on modelling and simulation: towards optimization of bread baking. Food Control 27:45–52. https://doi.org/10.1016/j.foodcont.2012.02.034
15. Mondal A, Datta AK (2008) Bread baking—a review. J Food Eng 86:465–474. https://doi.org/10.1016/j.jfoodeng.2007.11.014
16. Giannou V, Kessoglou V, Tzia C (2003) Quality and safety characteristics of bread made from frozen dough. Trends Food Sci Technol 14:99–108. https://doi.org/10.1016/S0924-2244(02)00278-9
17. Bredariol P, Vanin FM (2022) Bread baking review: insight into technological aspects in order to preserve nutrition. Food Rev Int 38:651–668. https://doi.org/10.1080/87559129.2021.1878211
18. Cappelli A, Lupori L, Cini E (2021) Baking technology: a systematic review of machines and plants and their effect on final products, including improvement strategies. Trends Food Sci Technol 115:275–284. https://doi.org/10.1016/j.tifs.2021.06.048
19. Zhang G, Chatzifragkou A, Charalampopoulos D, Rodriguez-Garcia J (2023) Effect of defatted melon seed residue on dough development and bread quality. LWT 183:114892. https://doi.org/10.1016/j.lwt.2023.114892

Check for updates

Chapter 7

Color and Volume Measurement

Guoqiang Zhang

Abstract

The frozen dough technology has been widely applied in bread production, laying a stable way for producing fresh bread at any time. Despite its advantages, freezing treatment may affect final bread quality. Color and volume are important quality parameters for bread products, which greatly influence consumer acceptance and purchase intention. Therefore, measuring these parameters is important to evaluate the final product quality and meet consumer expectations. This chapter outlines clear procedures for measuring bread color and volume, including the use of a handheld chrome meter for color measurement and a Volscan Profiler for volume measurement. These methods provide reliable and reproducible assessments of bread quality, helping other researchers to optimization of breadmaking processes and bread quality.

Keywords Color measurement, Bread volume, Specific volume, Bread physical parameter, Bread quality, Bakery products

1 Introduction

Bread is one of widely consumed staple in the world, providing significant nutrients in the daily dietary and forming an important part of a balanced diet [1]. With the increasing market demand, frozen dough technology has been introduced in breadmaking to reduce production burden and extend product shelf life [2, 3]. However, prolonged freezing period may adversely affect gluten network and yeast activity, leading to a deterioration in the quality of frozen dough bread [4, 5]. To this end, evaluation of the final bread quality is essential to provide consumer expect products with desirable quality.

Color is a key quality parameter in the bread products [6]. Many studies have demonstrated that color plays an important role in shaping expectations of food taste and acceptance, since consumers use color to provide an indication of freshness, flavor, and texture qualities [7, 8]. In the case of white bread, consumers expect a yellow-gold color on the crust [9]. During baking, the formation

Guoqiang Zhang and Jinying Guo (eds.), *Frozen Dough and Bread Quality*, Methods and Protocols in Food Science, https://doi.org/10.1007/978-1-0716-5206-0_7,

of color in bread products results from the Maillard reaction and caramelization, which are associated with many factors, such as processing methods, ingredient used, and baking temperature and time [10–12]. Therefore, measuring bread color is important to maintain bread quality and meet consumer expectation. Bread color can be measured using handheld color meter to obtain CIELAB coordinates (L*, a*, and b*) [13]. In the CIELAB coordinates, L* indicates lightness, ranging from 0 (black) to 100 (white), a* is the red/green coordinate (i.e., −a* = green and a* = red), and b* is the blue/yellow coordinate (i.e., −b* = blue and b* = yellow) [14].

In addition to bread color, volume is another important parameter for bread quality, which greatly influence consumer acceptance [15]. In general, high volume is considered a high quality characteristic of bread and is more attractive to consumers [16]. Additionally, it has been reported that bread volume is related to bread texture, with a decrease in bread volume generally increasing bread hardness, which has strong negative correlation [17, 18]. In order to maintain bread quality, measuring bread volume is important. Traditionally, the rapeseed displacement method has been widely used in determination of bread volume. However, this method has some drawbacks, such as operator dependence, tedious and laborious, and the potential crushing of soft products during operation, which affects measurement efficiency and the accuracy [19]. Volscan Profiler method (10-16.01) is an approved method by Cereals & Grains Association (AACC) in determination of bread volume [20]. The Volscan Profiler is a laser scanner that can automatically measure the volume and dimensional profile of bakery products, which can apply for various bakery products (e.g., loaf, muffins, and cakes), as well as products with challenging shapes [19].

To this end, this chapter describes procedures for measuring the color of bread crust and crumb with using a handheld color meter (Chroma meter 400, Konica Minolta Japan), as well as for measuring bread volume using Volscan Profiler method (10-16.01).

2 Materials

2.1 Color Measurement

1. Chroma meter (CR-400, Konica Minolta, Japan).

2.2 Volume Measurement

1. Balance.
2. Volscan Profiler C (Volscan 300, Stable Micro Systems, UK) (*see* **Note 1**).

3 Methods

3.1 Color Measurement

Bread color measurement will be conducted after the bread cooled to room temperature.

1. Calibrate the chroma meter using specific white calibration plate.
2. After calibration, place the chroma meter on the bread crust to measure crust color, and analyze three different points (Fig. 1A).
3. Place the chroma meter on the bread slice to measure crumb color and analyze three different points (Fig. 1B).

3.2 Volume Measurement

Bread volume measurement will be conducted after the bread cooled to room temperature. The procedure is illustrated in Fig. 2.

1. Weigh bread in a balance and record the weight of the sample (*see* **Note 2**).
2. Open New Scan software and input bread information, including sample name, data, weight, operation conditions (vertical step size of 10 mm and rotation speed of 1 rps), and then calibrate the height (*see* **Note 3**).
3. Open the scanner door, place the bread centrally and vertically onto the spikes of the support platform, and secure the sample using the supporting top pin (*see* **Notes 4, 5, and 6**).
4. Close the scanner door and then for measurement (*see* **Notes 1, 7, and 8**).
5. The results will be exported as 2D and 3D.

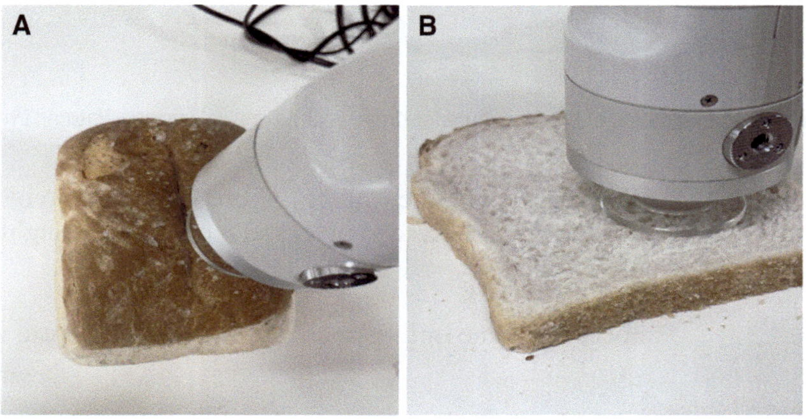

Fig. 1 Color measurement of bread; **(A)** crust color measurement; **(B)** crumb color measurement

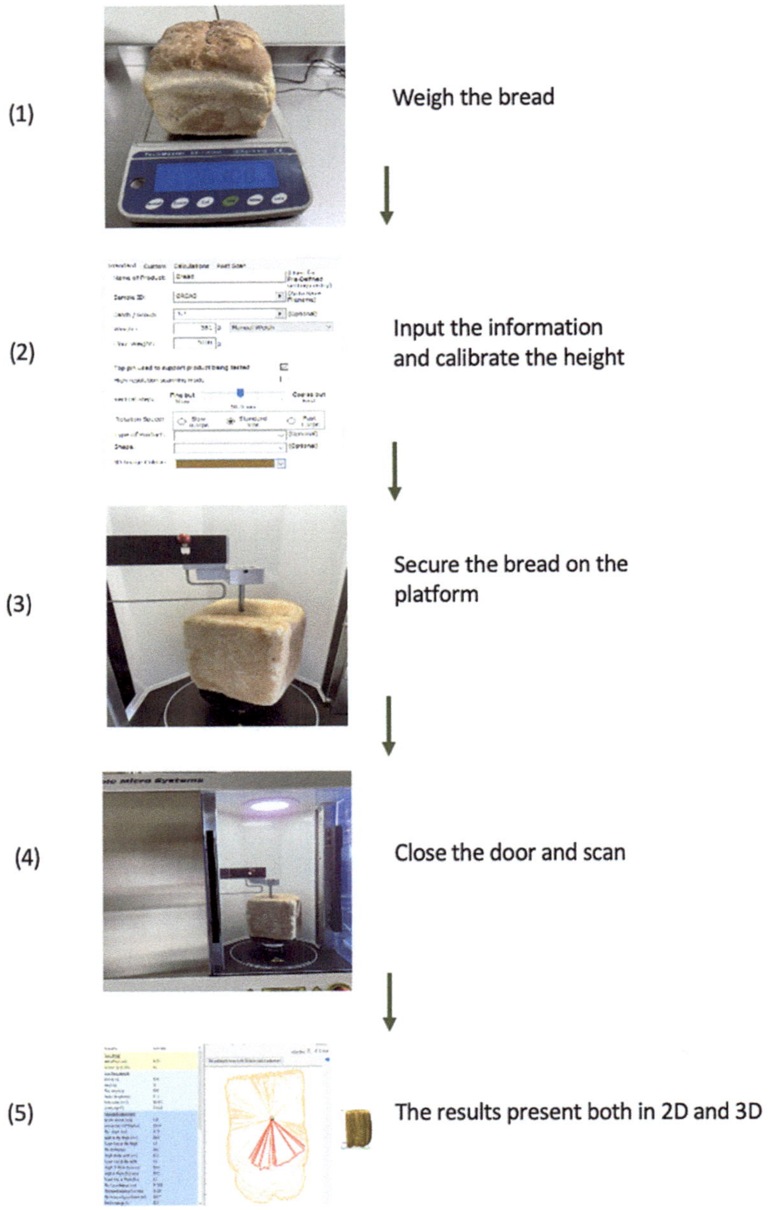

Fig. 2 Schematic representation of the procedure for bread volume analysis using a Volscan Profiler

The specific volume (mL/g) of bread is the ratio of the volume and mass of the bread sample, and is calculated using the following equation:

$$\text{Specific volume (mL/g)} = \text{bread volume (mL)}/\text{bread weight (g)}$$

4 Notes

1. The Volscan Profiler C uses a laser. Therefore, for safety reasons, do not deliberately look into the laser beam during operation.
2. Make sure the bread has cooled to room temperature before weighing.
3. During height calibration, ensure the machine door close with no product inside.
4. The spikes on the platform are sharp, please carefully secure the sample during operation.
5. Move the top arm support to allow space for the sample and this depends on the size and rigidity of the sample.
6. Avoid off center loading and gaps under the sample.
7. Before measurement, ensure the sample is located tight.
8. The centrifugal force generated during sample scanning may cause sample to fragment or wobble, particularly in the case of delicate samples. Therefore, a slower rotational speed is recommended. In addition, for more intricate products where fine details are necessary for accurate volume, a smaller vertical step size should be used.

References

1. Ribet L, Kassis A, Jacquier E, Monnet C, Durand-Dubief M, Bosco N (2025) The nutritional contribution and relationship with health of bread consumption: a narrative review. Crit Rev Food Sci Nutr 65:5698–5725. https://doi.org/10.1080/10408398.2024.2428593
2. He N, Xia M, Zhang X, He M, Li L, Li B (2024) Quality attributes and functional properties of whole wheat bread baked from frozen dough with the addition of enzymes and hydrocolloids. J Sci Food Agric 104:1928–1941. https://doi.org/10.1002/jsfa.13077
3. Jiang-Peng S, Jia-Qin F, Chuang L, Shou-Ping L, Chin-Ping T, Ping-Ping W, Xiong F, Chun C (2025) Alleviative effects of *Dendrobium officinale* polysaccharide on the quality deterioration of frozen dough and corresponding bread. Int J Biol Macromol 304:140705. https://doi.org/10.1016/j.ijbiomac.2025.140705
4. Xu K, Chi C, She Z, Liu X, Zhang Y, Wang H, Zhang H (2022) Understanding how starch constituent in frozen dough following freezing-thawing treatment affected quality of steamed bread. Food Chem 366:130614. https://doi.org/10.1016/j.foodchem.2021.130614
5. Jiang Z-J, Guo X-N, Xing J-J, Zhu K-X (2023) Alleviative effects of chitooligosaccharides on the quality deterioration of frozen dough subjected to freeze–thaw cycles. Food Hydrocoll 144:109016. https://doi.org/10.1016/j.foodhyd.2023.109016
6. Torbica A, Škrobot D, Janić Hajnal E, Belović M, Zhang N (2019) Sensory and physico-chemical properties of wholegrain wheat bread prepared with selected food by-products. LWT 114:108414. https://doi.org/10.1016/j.lwt.2019.108414
7. Veflen N, Velasco C, Kraggerud H (2023) Signalling taste through packaging: the effects of shape and colour on consumers' perceptions of cheeses. Food Qual Prefer 104:104742. https://doi.org/10.1016/j.foodqual.2022.104742
8. Chen W, Wu X, Liu Z, Liu Y, Liu Q, Pointer MR, Liang J, Khanh TQ (2022) The impact of illuminance level, correlated colour temperature and viewing background on the purchase intention for bread and cakes. Food Qual Prefer 98:104537. https://doi.org/10.1016/j.foodqual.2022.104537
9. Purlis E, Salvadori VO (2009) Modelling the browning of bread during baking. Food Res Int 42:865–870. https://doi.org/10.1016/j.foodres.2009.03.007

10. Zhang G, Li Z (2024) Impact of melon seed oil cake with different particle sizes on bread quality. Food Prod Process Nutr 6:44. https://doi.org/10.1186/s43014-024-00225-9
11. Purlis E (2010) Browning development in bakery products—a review. J Food Eng 99:239–249. https://doi.org/10.1016/j.jfoodeng.2010.03.008
12. Zhang G, Wang Y, Liang Z, Guo J, Wu B (2025) Nutritional value and advances of mealworm (*Tenebrio molitor*) as a sustainable functional ingredient in bakery products: a key step toward sustainable diets. Future Foods 12:100716. https://doi.org/10.1016/j.fufo.2025.100716
13. Huang S, Miskelly D (2019) Steamed bread—a review of manufacturing, flour quality requirements, and quality evaluation. Cereal Chem 96:8–22. https://doi.org/10.1002/cche.10096
14. Pathare PB, Opara UL, Al-Said FA-J (2013) Colour measurement and analysis in fresh and processed foods: a review. Food Bioproc Tech 6:36–60. https://doi.org/10.1007/s11947-012-0867-9
15. Pang J, Guan E, Yang Y, Li M, Bian K (2021) Effects of wheat flour particle size on flour physicochemical properties and steamed bread quality. Food Sci Nutr 9:4691–4700. https://doi.org/10.1002/fsn3.2008
16. Jiang X, Wang X, Zhou S (2022) Effect of flaxseed marc flour on high-yield wheat bread production: comparison in baking, staling, antioxidant and digestion properties. LWT 169:113979. https://doi.org/10.1016/j.lwt.2022.113979
17. Zhang G, Chatzifragkou A, Charalampopoulos D, Rodriguez-Garcia J (2023) Effect of defatted melon seed residue on dough development and bread quality. LWT 183:114892. https://doi.org/10.1016/j.lwt.2023.114892
18. Xin T, Tang S, Su T, Huang Z, Huang F, Zhang R, Dong L, Deng M, Shen Y, Su D (2022) Impact of replacing wheat flour with lychee juice by-products on bread quality characteristics and microstructure. LWT 165:113696. https://doi.org/10.1016/j.lwt.2022.113696
19. de Beer T (2023) Quality assessment of wheat flour, dough, and bread. ICC Handbook of 21st Century Cereal Science and Technology. Academic Press, pp 225–233. https://doi.org/10.1016/B978-0-323-95295-8.00002-2
20. Smewing J (2016) AACCI approved methods Technical Committee Report on the Collaborative Study for a New AACCI Method (10-16.01): volumetric and dimensional profile determination of baked products using laser topography—VolScan Profiler method. Cereal Foods World 61:18–23. https://doi.org/10.1094/CFW-61-1-0018

Chapter 8

Texture Profile Analysis of Bread

Chunyan Wang and Rina Wu

Abstract

Bread has been widely appreciated as a staple food for providing of important nutrients. Texture is a key characteristic of bread, offering insights into its quality, ingredients, and baking methods. The assessment of product quality through texture analysis is an essential tool for both bread products development and production. Texture profile analysis (TPA), by simulating the mechanical effects during the chewing process, provides a quantitative, repeatable, and accurate method for determining the physical properties of bread, including hardness (firmness), springiness, resilience, and cohesiveness. This chapter provide a clear procedure for the TPA determination of bread, aiming to facilitate the reader understanding of the TPA measurement steps. The method specifies the preparation of samples, machine parameters, and data calculation method to be used. And the slices of bread are taken as an example.

Keywords Bread, Slice, Texture analysis, Hardness, Springiness, Tesilience, Cohesiveness

1 Introduction

Bread is a well-known, principal staple food all over the world especially in developing countries due to its economic, nutritional, and satisfactory importance [1]. Bread is a bakery item typically produced using wheat flour, water, salt, yeast, and dough improvers. These components are combined in specific proportions to form a viscoelastic structure in the dough, which is finally baked to produce the finished product [2]. Both the final product appearance and its texture are affected significantly by the bread recipes and methods employed in dough processing combined with the conditions of baking. The texture a key characteristic of bread because of its link to eating qualities, flavor, shelf life, and consumer acceptance, and the variations in bread textural quality are still of concerns for the bread making industry [3–5]. The texture analysis data unlocks a deeper understanding of quality attributes, which is instrumental in optimizing both products and processes, as well as in guiding new product development.

Guoqiang Zhang and Jinying Guo (eds.), *Frozen Dough and Bread Quality*, Methods and Protocols in Food Science,
https://doi.org/10.1007/978-1-0716-5206-0_8,

The texture can be measured objectively (instrumentally) and subjectively (sensorially). Instrumental test simulates chewing conditions and presents significant correlation with sensory texture measures. Texture profile analysis (TPA), as an instrumental test method, provides quantifiable, repeatable, and accurate data on the physical properties of different foods by imitating the mechanical behavior of mastication [6]. TPA is a form of controlled compression test that will either subject the product to a standard force applied for a fixed time or compress the sample though a given distance and measure the force required to achieve a given percentage thickness compression [7]. Currently, the texture profile analysis has been widely used in bread measurement [8–11]. Figure 1 shows a typical TPA curve for breadcrumb from a double compression test. The determination of the texture of bread mainly includes hardness, springiness, cohesiveness, and resilience four important indicators.

Hardness (popular nomenclature: softness, firmness)—it is of primary concern and might be used to describe the crust and bread crumb. Hardness (firmness) can give an indication of the amount of physical "staling" has occurred in the product. It is defined as the maximum force, measured in Newtons (N), necessary to achieve a given deformation during the first compression cycle. This value, represented by the highest peak force recorded, correlates with the force required for the first bite [12].

Springiness and resilience—describe the way crumb springs back and returns to its undeformed state after the first compressing force has been removed. They are two important parameters to evaluate the staling process of bread during storage [7]. The springiness is calculated dividing the time (in seconds or millimeters) necessary to reach the highest point of the peak in the second compression and in the first one. The resilience is calculated by dividing the upstroke area of the first compression peak by the downstroke one.

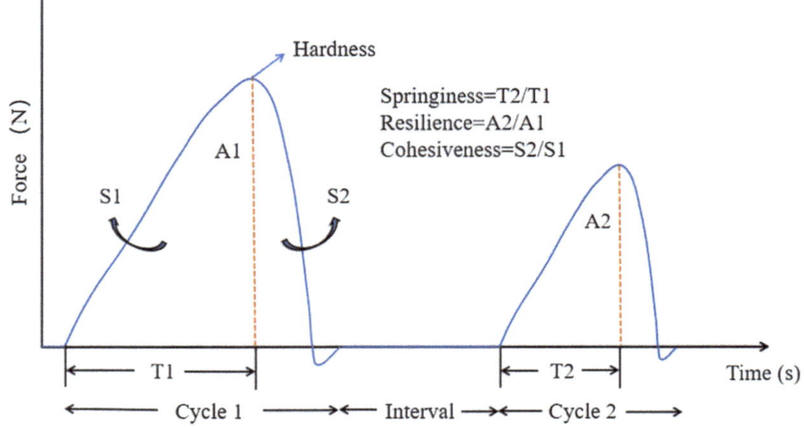

Fig. 1 Graphical results of a texture profile analysis (TPA) of bread crumb. The calculations of different parameters results are shown in figure

Cohesiveness—generally seen as a positive character in bread [7]. It represents how well the product withstands a second deformation taking into account its resistance during the first deformation. Cohesiveness is calculated by dividing areas of work during the second compression by that of the first one [12]. The manifestation of cohesiveness is determined by moisture content and the structural integrity of the crumb. The reduced moisture levels lead to an increase in crumbliness.

This chapter outline a clear procedure for the TPA determination of bread, aiming to facilitate the reader understanding of the analysis steps of TPA. The method specifies the preparation of samples, machine parameters, and data calculation method to be used. TPA could be performed either on the loaf or on the bread slice. In this chapter, we take slices of bread as an example.

2 Materials

2.1 Bread Samples

1. Bread loaf of 200 g.

2.2 Apparatus

1. Textural profile analyzer (TA-XT plus, Stable Micro Systems, Godalming, UK) with load cell of 10 kg (*see* **Note 1**).
2. Cylindrical probe of 36 mm in diameter (P36) (*see* **Note 2**).
3. Bread knife or automatic bread slicer.
4. Plastic bags.
5. Room with temperature of approximately 25°C or incubator.

3 Methods

1. Preparation of sample
   Total sample thickness is 25 mm (*see* **Note 3**). Loaves may be sliced either mechanically or by hand, 25 mm thick, and discard heel slices of loaf (this is preferred considering the high reproducibility of the slice thickness) (*see* **Notes 4 and 5**). Use six central slices from each loaf to perform textural analysis. If the textural profile analysis is not performed within 4 h, put bread slice in plastic bags (not more than 24 h) before analysis.
2. Install the 36 mm cylindrical probe to the instrument and use the TPA analyzer software for setting the correct probe.
3. Set the zero point between the plate and the plunger.
4. Set the analysis for the double compression test.

5. Position upper crosshead limit so that compression plunger is 1 mm above center surface of sample.
6. Position lower crosshead limit at 40% compression (10 mm compression depth).
7. Set crosshead speed (rate of compression) at 1 mm/s (*see* **Note 6**).
8. Set the interval at 5 s between the first and second compressions.
9. Set chart speed at 5 mm/s (5:1 ratio of chart to crosshead extension).
10. Put the sample on the center of the plate. Be careful to evaluate the position of the compression plunger, avoiding any irregular or not representative areas of bread slices.
11. Start the analysis.
12. Get the results and assess the manner in which they are presented (*see* **Note 7**).
13. For firmness evaluation, see the force reading at 25% compression, which corresponds to the force registered after 31 mm from the start of the force recording.
14. Return compression plunger to upper limit position.
15. Discard test slice.
16. Repeat Steps 10~15 until desired number of slices have been evaluated. Evaluate six slices per loaf.

3.1 Calculation

1. The compression force value of all readings per slice (or slices) may be reported in one of the following ways.
   a. in kilograms of force. This value is read directly off chart.
   b. in newtons of force. Many testing machines may be calibrated directly in newtons (N), an SI unit. If this capability is unavailable, conversion is as follows: N = kilogram × (9.8 m/s^2)

      $$\text{Example : kg of force} = 0.3$$

      $$(0.3)(9.8\ \text{m/sec}^2) = 2.94\text{N}$$

2. Stress can be measured indirectly by converting the compression force value (measured in newtons) to newtons of force/unit area of plunger = N/ (1.0179×10^{-3} m^2) (for 36-mm diameter probe).

   $$\text{Example : } 2.94\text{N}/1.0179 \times 10^{-3}\text{m}^2 = 2888.3\ \text{N}/m^2$$

4 Notes

1. The load cell is an important parameter to evaluate before performing the TPA. In order to make the optimal selection, it is crucial to determine the greatest force that may be reached during the analysis. Indeed, the evaluation of textural properties of the bread slice implies lower force compared to breadcrust evaluation, and a 10 kg load cell (or 100 N load cell) might be the best option. The proper choice of load cells could increase the accuracy of the test.
2. The typical probe for bread evaluation is aluminum cylinder with a diameter of 36 mm. However, other cylindrical probes with a diameter of 30~40 mm could also be used, if it is not possible to use the standard one.
3. Total sample thickness is 25 mm; therefore, for evaluating commercially available sliced bread, put two or three slices stacked together. Always use the central slice of the loaf.
4. The storage, packaging, and handling of samples prior to analysis are variable factors in bread testing. It is important to identify these conditions and keep them consistent within a lab to ensure comparable findings.
5. Cutting crust off the bread immediately before the compression test is also considered a variable testing procedure. For central bread slice from 200 g loaf of soft bread, the crust does not interfere with compression by a 36-mm diameter plunger. However, if a larger plunger or a different type of bread is used so that crust does resist compression, the crust may be removed from the bread sample before testing.
6. A speed of 100 mm/min is used in the official method, which corresponds approximately at 1.7 mm/s. However, a speed of 2 mm/s could also be used.
7. Some software report the results as the force registered during the analysis; other software report them related to a surface of 1 m^2 (calculating the surface area of the plumber).

References

1. Mollakhalili-meybodi N et al (2023) Sensory attributes of wheat bread: a review of influential factors. J Food Meas Charact 17(3):2172–2181
2. Canale M et al (2025) Utilization of wheat malts as a replacement for improvers to enhance physicochemical, technological, nutritional, and sensory characteristics of bread. LWT 219:117535
3. Young LS (2012) 22—Applications of texture analysis to dough and bread. In: Cauvain SP (ed) In breadmaking, 2nd edn. Woodhead Publishing, Cambridge, pp 562–579
4. Estévez-López RD et al (2021) Influence of bread shape on the sensory characteristics of Galician breads: development of lexicon, efficacy control of the trained panel and establishment of a sensory profile. LWT 135:110024
5. Dong Y, Karboune S (2021) A review of bread qualities and current strategies for bread

bioprotection: flavor, sensory, rheological, and textural attributes. Compr Rev Food Sci Food Saf 20(2):1937–1981
6. Saldaña E et al (2015) Microstructure, texture profile and descriptive analysis of texture for traditional and light mortadella. Food Struct 6:13–20
7. Cauvain SP (2020) Chapter 17—applications of rheological analysis to dough and texture analysis to bread. In: Cauvain SP (ed) Breadmaking, 3rd edn. Woodhead Publishing, Cambridge, pp 517–540
8. Tagliasco M et al (2025) The effects of gluten addition, dough moisture content and mixing time on the textural properties and in vitro starch digestibility of durum wheat bread made with coarse semolina. Appl Food Res 5(1):101017
9. Corrado M et al (2023) Impact of storage on starch digestibility and texture of a high-amylose wheat bread. Food Hydrocoll 135:108139
10. Ruderman M et al (2025) Digital image analysis to assess the texture of bread products. Appl Food Res 5(2):101447
11. Rahman MS et al (2021) Measurement of instrumental texture profile analysis (TPA) of foods. In: Khan MS, Shafiur Rahman M (eds) Techniques to measure food safety and quality: microbial, chemical, and sensory. Springer International Publishing, Berlin, pp 427–465
12. Montemurro M, Pontonio E (2024) Texture profile analysis. In basic. In: Gobbetti M, Rizzello CG (eds) Methods and protocols on sourdough. Springer, New York, pp 103–110

Chapter 9

Proximate Composition Analysis

Yuanqing Luo, Zhao Qin, Xiaoshuang Cai and Hua-Min Liu

Abstract

Frozen dough and related bread products hold significant importance in the global food market. Proximate composition analysis provides essential insights into the chemical makeup of these products and plays a critical role in product development and quality control. This chapter systematically details the standardized procedures for determining moisture content using air-oven and vacuum-oven methods, ash content through dry ashing, protein content via Kjeldahl and Dumas methods, crude fat content by Soxhlet extraction, and dietary fiber content employing the enzymatic-gravimetric method. These methods facilitate a better understanding of how each component influences the physicochemical properties of frozen dough and the sensory characteristics of bread, thereby aiding in the optimization of bread-making processes and the improvement of overall product quality.

Keywords Proximate analysis, Moisture determination, Ash content, Protein content, Crude fat, Dietary fiber

1 Introduction

Proximate composition typically refers to the five fundamental categories of components in food: moisture, protein, fat, carbohydrates (including dietary fiber), and ash. These components serve as essential data for nutritional evaluation, quality control, and process design. Accurate proximate composition analysis is crucial for understanding and improving the quality of frozen dough and bread [1].

Moisture is one of the most critical components in frozen dough, as its content directly affects the physicochemical properties of the dough and the quality of the final bread. Research indicates that moisture migration and recrystallization during the freezing process can influence the structural stability of the dough [2–4]. Specifically, excessive moisture can lead to the formation of large ice crystals that disrupt the gluten network, resulting in reduced loaf volume and a rough texture. Therefore, precise measurement of moisture content is vital for optimizing freezing process parameters, predicting changes in dough quality, and enhancing the sensory qualities of bread.

Guoqiang Zhang and Jinying Guo (eds.), *Frozen Dough and Bread Quality*, Methods and Protocols in Food Science,
https://doi.org/10.1007/978-1-0716-5206-0_9,

The air-oven method and vacuum-oven method are two commonly employed techniques for moisture content determination [5, 6].

Ash content primarily reflects the degree of milling precision of flour (e.g., roughly 1.5% ash in whole wheat flour compared to about 0.5% in refined flour) and the mineral content (such as calcium and iron) [7, 8]. Excessive ash content can result in bread with a darker color and a coarser texture, whereas insufficient ash may adversely affect dough fermentation [9]. Measuring ash content can monitor the quality of flour raw materials, ensuring consistency in frozen dough. Dry ashing is a common method for ash analysis, which involves burning the sample at high temperatures until all organic materials are eliminated, leaving only the inorganic mineral portion [10]. This method is straightforward and widely applied in the analysis of frozen dough composition.

Protein is a key component that contributes to the structural formation of dough, with its content and properties directly influencing the texture and mouthfeel of bread. Freezing can lead to protein denaturation (such as disulfide bond breakage), which diminishes gluten activity [11]. By determining total protein content using the Kjeldahl or Dumas methods, and analyzing protein composition with techniques like SDS-PAGE, one can assess the impact of freezing on protein structure, providing guidance for selecting freeze-resistant proteins or adjusting flour formulations [12, 13]. The Kjeldahl method, though complicated in its procedure, is a classic method known for its stable and reliable results. In recent years, the Dumas method has gained popularity in the analysis of cereal protein due to its speed and accuracy [12, 14].

Fat is a key functional component in frozen dough and bread, with its level and physical state regulating dough processing properties (e.g., extensibility and moldability) and influencing the sensory quality and storage stability of the final product [15]. Excessive fat compromises dough handling and freezing stability, leading to weakened bread structure. Insufficient fat fails to provide adequate lubrication and antistaling effects, leading to increased dough toughness, reduced fermentation capacity, and a coarse, easily hardened bread texture [16]. Precise fat determination is thus critical for quality control. The Soxhlet extraction method remains the industry standard for frozen dough and bread due to its simplicity, cost-effectiveness, and consistent reliability.

Dietary fiber is an important functional component in bread, and its content and type significantly impact the nutritional value and sensory characteristics of the product. Incorporating ingredients rich in dietary fiber can enhance the texture of frozen dough, minimize water migration, and help prevent deterioration of the gluten network during storage [17]. However, excessive addition can disrupt the gluten network [18]. The enzymatic gravimetric method is currently the most authoritative method for measuring dietary fiber, as it uses specific enzyme systems to breakdown nonfiber components before determining fiber content. This method can distinguish between total dietary fiber, soluble dietary fiber, and insoluble dietary fiber, providing a scientific basis for the development of functional bread.

This chapter aims to systematically introduce key technologies for analyzing the proximate composition (moisture, protein, ash, fat, and dietary fiber) of frozen dough and bread. Standardized operating procedures are outlined to provide comprehensive analytical guidance for baking enterprises and research institutions, supporting the industrial upgrading of frozen dough technology and the development of healthier bread products. These methods are derived from scientific literature, international standards, and practical experience.

2 Materials

2.1 Materials for Moisture Content Determination

2.1.1 Air-Oven Methods

Equipment and Supplies:

1. Analytical balance, accurate to ±0.0001 g.
2. Sample pulverizing equipment (e.g., grinding mill and homogenizer).
3. Flat aluminum or glass weighing dishes or bottles with sealing caps (*see* **Note 1**).
4. Electric constant-temperature drying oven (capable of maintaining 103 ± 2°C)
5. Desiccator containing an effective desiccant (*see* **Note 2**).

2.1.2 Vacuum-Oven Method (Suitable for High-Sugar Bread)

Equipment and Supplies:

1. Analytical balance, accurate to ±0.0001 g.
2. Flat aluminum or glass weighing dishes or bottles with lids.
3. Vacuum drying oven (accuracy: ±2°C, vacuum level ≤ 100 mmHg [13.3 kPa]).
4. Vacuum pump (capable of maintaining stable pressure).
5. Desiccator, containing an effective desiccant.

2.2 Materials for Ash Content Determination

2.2.1 Dry Ashing Method

Equipment and Supplies:

1. Analytical balance, accurate to ±0.0001 g.

2. High-temperature muffle furnace, equipped with an automatic pyrometer, and capable of operating at temperatures up to 600 ± 15°C.
3. Porcelain crucibles or beakers made of platinum or silica, with a capacity of 100–200 mL.
4. Desiccator, containing an effective desiccant.

2.3 Materials for Protein Content Determination

2.3.1 Kjeldahl Method

Equipment and Supplies:

1. Kjeldahl nitrogen determination apparatus (semiautomated or fully-automated).
2. Digestion unit with temperature control range from ambient to 500°C.
3. Analytical balance, accurate to ±0.0001 g.
4. Digestion tubes or digestion flasks (250 mL).
5. Pipettes (10 mL and 25 mL).

Chemicals and Reagents:

1. General requirements: all reagents shall be recognized analytical grade and nitrogen-free, unless otherwise specified. Use distilled, deionized water, or water of equivalent purity.
2. Specific reagents: concentrated sulfuric acid (H_2SO_4, ≥98%), copper(II) sulfate ($CuSO_4$), potassium sulfate (K_2SO_4), sodium hydroxide (NaOH), boric acid (H_3BO_3), methyl red, methylene blue or bromocresol green, 95% ethanol (C_2H_5OH), hydrochloric acid (HCl), and sucrose (optional, for method validation).
3. Reagent preparation

    1. NaOH solution, 40% (w/v): dissolve 40 g NaOH in water and dilute to 100 mL.
    2. Indicator solutions (*see* **Note 3**):
    Indicator solution A: mix 2 g/L methyl red solution in 95% ethanol and 2 g/L methylene blue solution in 95% ethanol in a 2:1 (v/v) ratio.
    Indicator solution B: mix 2 g/L methyl red solution in 95% ethanol and 2 g/L bromocresol green solution in 95% ethanol in a 1:5 (v/v) ratio.
    3. Boric acid indicator solution, 2% (w/v): dissolve 20 g H_3BO_3 in approximately 800 mL of warm water. Cool, add 10 mL of either Indicator Solution A or B (prepared above), and dilute to 1 L with deionized water.

4. Standard acid titrant: 0.05 mol/L H_2SO_4 standard solution, or 0.1 mol/L HCl standard solution (*see* **Note 4**).

2.3.2 Dumas Method

Equipment and Supplies:

1. Dumas nitrogen/protein analyzer: equipped with a thermal conductivity detector (TCD) and capable of measuring nitrogen content in the range of 0.1–200 mg. The instrument should have calibration capabilities.
2. Analytical balance, capable of weighing to an accuracy of ±1 mg.
3. Tin foil squares, for encapsulating samples.
4. Forceps, for handling tin foil squares and samples.

Chemicals and Reagents:

1. Combustion gas (O_2), purity ≥99.995%.
2. Carrier gas (CO_2, He, or Ar), purity ≥99.995%.
3. Nitrogen standard reference materials: Aspartic acid or Urea, purity ≥99.99%, for instrument calibration and validation.

2.4 Materials for Crude Fat Determination

2.4.1 Soxhlet Extraction Method

Equipment and Supplies:

1. Soxhlet extractor (150 or 250 mL), consisting of an extraction flask, extraction chamber, and condenser (*see* **Note 5**).
2. Analytical balance, capable of weighing to an accuracy of ±0.001 g.
3. Electric constant-temperature drying oven, capable of maintaining a temperature of 103 ± 2°C.
4. Water bath with adjustable temperature control between 60°C and 80°C.
5. Nitrogen blowdown apparatus (optional), to accelerate removal of residual petroleum ether and prevent oxidation of the fat.
6. Desiccator containing an effective desiccant, such as indicating silica gel.
7. Defatted filter paper for preparing sample cartridges.

Chemicals and Reagents:

1. Petroleum ether, boiling range 30–60°C (*see* **Note 6**).

2.5 Materials for Dietary Fiber Determination

2.5.1 Enzymatic-Gravimetric Method

Equipment and Supplies:

1. High-speed grinder.
2. Analytical balance, capable of weighing to an accuracy of ±0.0001 g.
3. Tall-form beakers (400 or 600 mL).
4. Temperature-controlled shaking water bath:
    1. Operating range: Ambient +5°C to 100°C
    2. Temperature fluctuation: ±1°C
5. pH meter with temperature compensation (±0.1 unit accuracy). Calibrate before use with standard buffer solutions at pH 4.0, 7.0, and 10.0.
6. Vacuum filtration system:
    1. Vacuum pump with regulator or aspirator.
    2. 1 L thick-walled filtration flask, with sidearm for vacuum connection.
    3. Rubber adapter for flask-funnel seal.
7. Filtering crucibles: glass filter plate with pore size 40–60 μm. Pretreatment process:
    1. Burn cleaned crucibles at 550 ± 25°C in a muffle furnace for 6 h. Remove crucibles when furnace temperature drops below 130°C.
    2. Use vacuum apparatus to remove diatomaceous earth and ash.
    3. Soak at room temperature in potassium dichromate cleaning solution for 1 h.
    4. Rinse crucibles with water followed by deionized water, then use 15 mL acetone for rinsing and air dry.
    5. Add 1.0 g celite to dry crucible, dry at 130 ± 3°C to constant weight.
    6. Cool crucible in desiccator for 1 h, weigh and record the mass of the crucible (including celite, m_{CC}), precise to 0.1 mg.
8. Muffle furnace: temperature controllable to 550 ± 25°C, with accuracy ±5°C.
9. Oven: temperature controllable to 105 ± 2°C, 130 ± 3°C.

10. Desiccator: containing silica gel or equivalent desiccant. Dry desiccant overnight in a 130°C oven every 2 weeks.
11. Supporting items: pipettes and tips (100 μL and 5 mL), graduated cylinders, magnetic stirrer with bars, rubber-tipped spatulas.

Chemicals and Reagents:

Use deionized water throughout the procedure. Reagents are analytical grade unless otherwise specified.

1. Ethanol solutions (78% v/v and 95% v/v).
2. Acetone.
3. HCl solution (0.561 M).
4. NaOH solution (1 M).
5. MES-TRIS buffer (0.05 M each, pH 8.2 at 24°C):
    1. Dissolve 19.52 g MES [2-(N-morpholino) ethanesulfonic acid] and 12.2 g TRIS [tris(hydroxymethyl)aminomethane] in 1.7 L water.
    2. Adjust to pH 8.2 (at 24°C) using 6 M NaOH or 4 M HCl, make volume up to 2 L with water (*see* **Note 7**).
6. Enzymes:
    1. Thermostable α-amylase (≥3200 μ/mL).
    2. Protease (≥350 μ/mL).
    3. Amyloglucosidase (≥3300 μ/mL).
7. Potassium dichromate ($K_2Cr_2O_7$) cleaning solution:
    1. Dissolve 100 g $K_2Cr_2O_7$ in 200 mL water, cool in an ice bath, then slowly add 1800 mL concentrated H_2SO_4 with constant stirring to prevent splashing.
8. Diatomaceous earth (acid-washed, preburned).

3 Methods

3.1 Moisture Content Determination [5, 6, 19]

3.1.1 Sample Preparation (see **Note 8**)

Frozen Dough:

1. Thaw frozen dough overnight in a 4°C refrigerator until it reaches a pliable, nonsticky consistency.

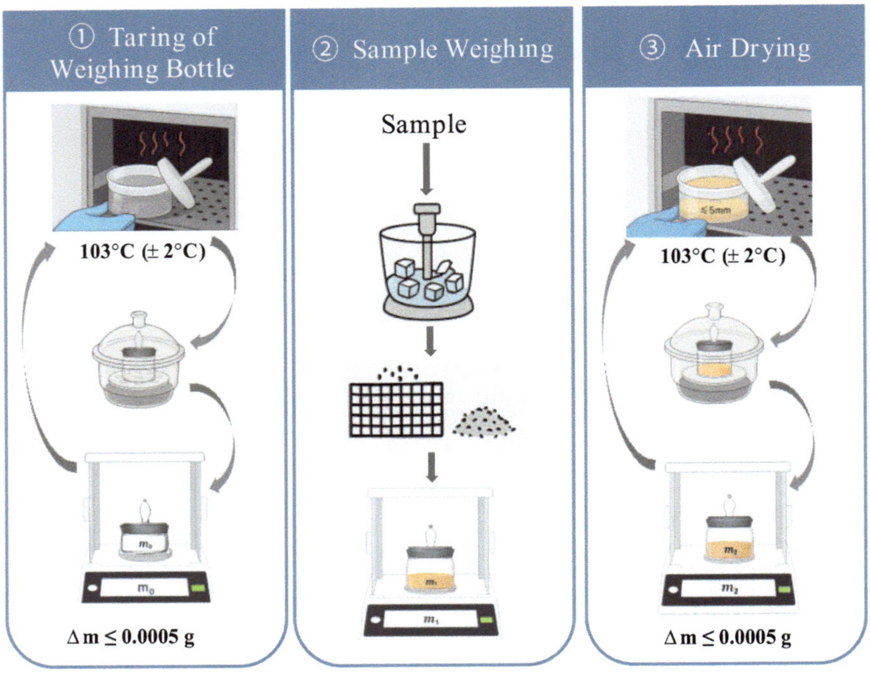

Fig. 1 Schematic diagram of moisture content determination by air-oven method

2. Dice the thawed dough and homogenize it at low speed with intermittent pulsing to form a uniform paste or granules (particle size ≤2 mm).

Bread:

1. Select representative bread pieces and cut them into small pieces. Process in a homogenizer to achieve a uniform particle size that passes through a 20-mesh sieve (aperture ~0.85 mm).
2. For samples that are difficult to mill, slice into 2–3 mm thick pieces.

3.1.2 Air-Oven Method

The procedure is illustrated in Fig. 1.

Taring of Weighing Bottle

1. Place an uncovered weighing bottle (with the lid alongside) in an oven at 103 ± 2°C for 1 h.
2. Cover the bottle and immediately transfer it to a desiccator to cool for 0.5 h. Weigh to an accuracy of ±0.0001 g.
3. Repeat drying in 1-h cycles until constant mass is achieved ($\Delta m \leq 0.0005$ g). Record tare mass (m_0) of weighing bottle.

Sample Weighing and Drying

4. Transfer approximately 2 g (±0.0001 g) of prepared sample into the tared bottle. Spread uniformly (layer thickness ≤5 mm), cover, and weigh. Record the mass (m_1) of bottle and nondried sample.
5. Place the weighing bottle (with the lid alongside) containing the samples in a drying oven at 103 ± 2°C. Air dry for 4 h (bread) or 5 h (frozen dough).
6. Cover the bottle, and immediately place it in a desiccator to cool for 0.5 h. Weigh (accuracy: ±0.0001 g).
7. Repeat drying in 1-h cycles until constant mass is achieved. Record the mass (m_2) of bottle and dried sample.

3.1.3 Vacuum-Oven Method (for Thermally Sensitive or High-Sugar Samples)

The procedure is illustrated in Fig. 2.

Taring of Weighing Bottle

1. Follow Steps 1–3 of the air-oven method (Sect. 3.1.2).

Sample Weighing

2. Transfer 2 g (± 0.0001 g) sample into the tared bottle. For high-sugar samples, add 0.5–1 g of celite, mix thoroughly to prevent sticking to the walls.
3. Spread the sample uniformly on the bottom of the weighing dish with a layer thickness ≤5 mm (to ensure complete evaporation of moisture). Cover, weigh, and record the mass (m_1) of bottle and nondried sample.

Vacuum Drying

4. Place the weighing bottle (with the lid alongside) containing the samples in the vacuum oven.
5. Connect the vacuum oven to a vacuum pump and evacuate slowly to 13.3–26.7 kPa (100–200 mm Hg), while simultaneously heating to 70 ± 2°C.
6. Maintain the vacuum and temperature for 4–6 h.
7. After drying, turn off the heating while maintaining vacuum for 10 min to allow the sample to cool to near room temperature.
8. Release vacuum slowly via the dry air inlet valve (*see* **Note 9**).
9. Open oven only after pressure has normalized.
10. Immediately cover the bottle, transfer it to a desiccator to cool for 0.5 h, and weigh.
11. Repeat drying in 1-h cycles until constant mass is achieved. Record the mass (m_2) of bottle and dried sample (*see* **Note 10**).

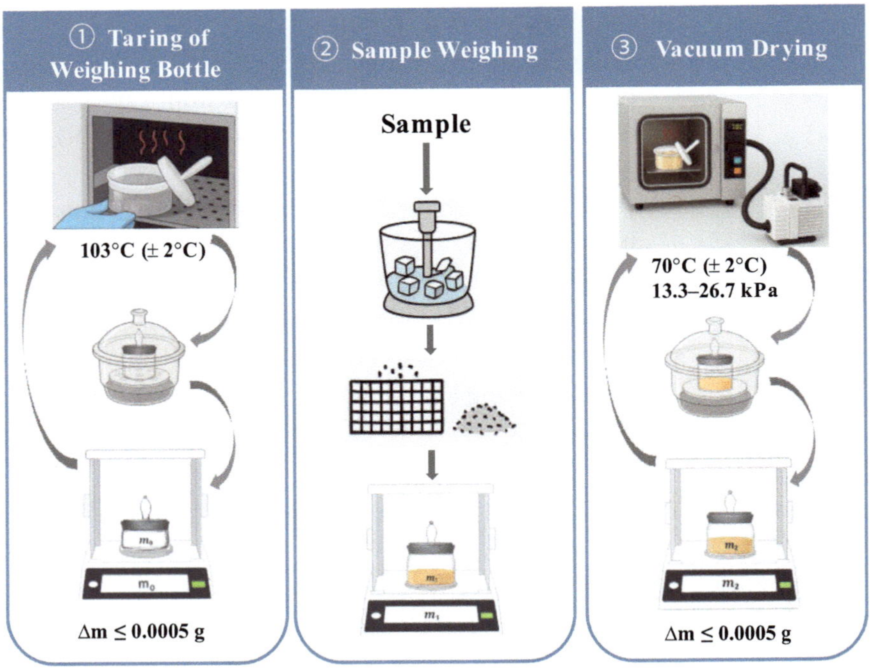

Fig. 2 Schematic diagram of moisture content determination by vacuum-oven method

3.1.4 Calculation

The moisture content (%) is calculated using Eq. (1) (*see* **Note 11**):

$$\%\mathbf{Moisture} = \frac{\mathbf{m_1 - m_2}}{\mathbf{m_1 - m_0}} \times \mathbf{100} \quad (1)$$

in which,
m_0 is the mass of the weighing bottle.
m_1 is the total mass of the bottle and nondried sample.
m_2 is the total mass of the bottle and dried sample.
Reporting: Express results to 0.1% precision.

3.2 Ash Content Determination [20–22]

Dry Ashing Method

The procedure is illustrated in Fig. 3.

3.2.1 Crucible Preparation and Taring

1. Cleaning:
   1. Soak porcelain crucibles in boiling dilute hydrochloric acid for 30 min.
   2. Rinse thoroughly with tap water followed by distilled water.

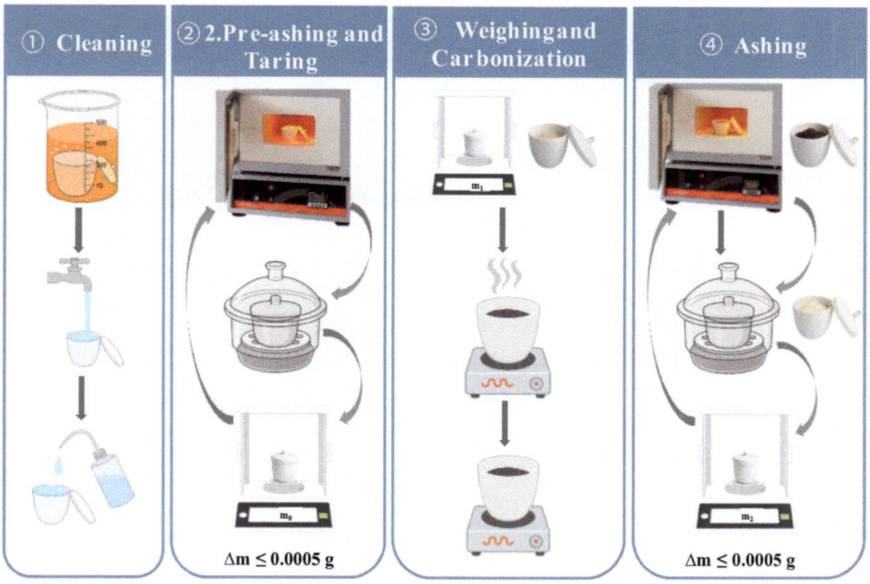

Fig. 3 Schematic diagram of ash content determination by dry ashing method

2. Preashing and Taring:

   1. Ash cleaned crucibles at 550 ± 25°C for 1 h.
   2. Remove and place in a desiccator to cool to room temperature. Weigh (accurate to 0.0001 g).
   3. Repeat ashing in 1-h cycles until constant mass is achieved. Record this weight as m_0.

3.2.2 Sample Preparation and Weighing

1. Prepare samples following the method outlined in Sect. 3.1.1 (grind to ≤0.85 mm).
2. Dry the pulverized sample to constant mass according to the air-oven method (Steps 4–7 of Sect. 3.1.2) (*see* **Note 12**).
3. Weigh 2–5 g (±0.0001 g) of the prepared sample into the tared crucible. Record this weight as m_1.

3.2.3 Carbonization

1. Place the crucible on an electric stove and heat slowly over low heat to carbonize the sample (produce black smoke) (*see* **Note 13**).
2. Stop heating when no more black smoke is emitted, and the sample appears dark-brown.

3.2.4 Ashing

1. Place the crucible containing the carbonized sample in a muffle furnace at 550 ± 25°C and ash it for 4 h (*see* **Note 14**).
2. Remove the crucible, cool in a desiccator for 30 min, and weigh.

3. Repeat ashing in 1-h cycles until constant mass is achieved. Record the mass (m_2) of the crucible and ash.

3.2.5 Calculation

Ash content (dry basis, %) is calculated using Eq. (2):

$$\%\mathbf{Ash} = \frac{\mathbf{m}_2 - \mathbf{m}_0}{\mathbf{m}_1 - \mathbf{m}_0} \times \mathbf{100} \quad (2)$$

where,
m_0 is the mass of the crucible.
m_1 is the total mass of the crucible and dried sample.
m_2 is the total mass of the crucible and ash.
Reporting: Express results to 0.01% precision.

3.3 Protein Content Determination [23–26]

3.3.1 Sample Preparation

1. Prepare samples following the method outlined in Sect. 3.1.1 (grind to ≤0.85 mm).
2. Dry the pulverized sample to constant mass according to the air-oven method (Steps 4–7 of Sect. 3.1.2) (*see* **Note 15**).

3.3.2 Kjeldahl Method

The procedure is illustrated in Fig. 4.

Digestion (*see* **Note 16**)

1. Place a 250 mL digestion tube in the digestion rack. Add 2 g of sample, 0.4 g of $CuSO_4$, and 6 g of K_2SO_4 (*see* **Note 17**).
2. Slowly add 20 mL concentrated H_2SO_4 to the digestion tube (*see* **Note 18**).
3. Place the digestion rack with tubes on the digester and increase the temperature slowly to 420°C. Digest the samples for at least 1 h until the digestion solution turns green and clear (no black particles) (*see* **Note 19**).
4. After digestion, turn off the digester, remove the rack, and cool the digested solution to room temperature before analyzing (*see* **Note 20**).

Automated Distillation and Titration

Optimize analysis parameters according to different instruments. Take the operating procedure of the Foss Kjeltec 8400 Kjeldahl Analyzer as an example.

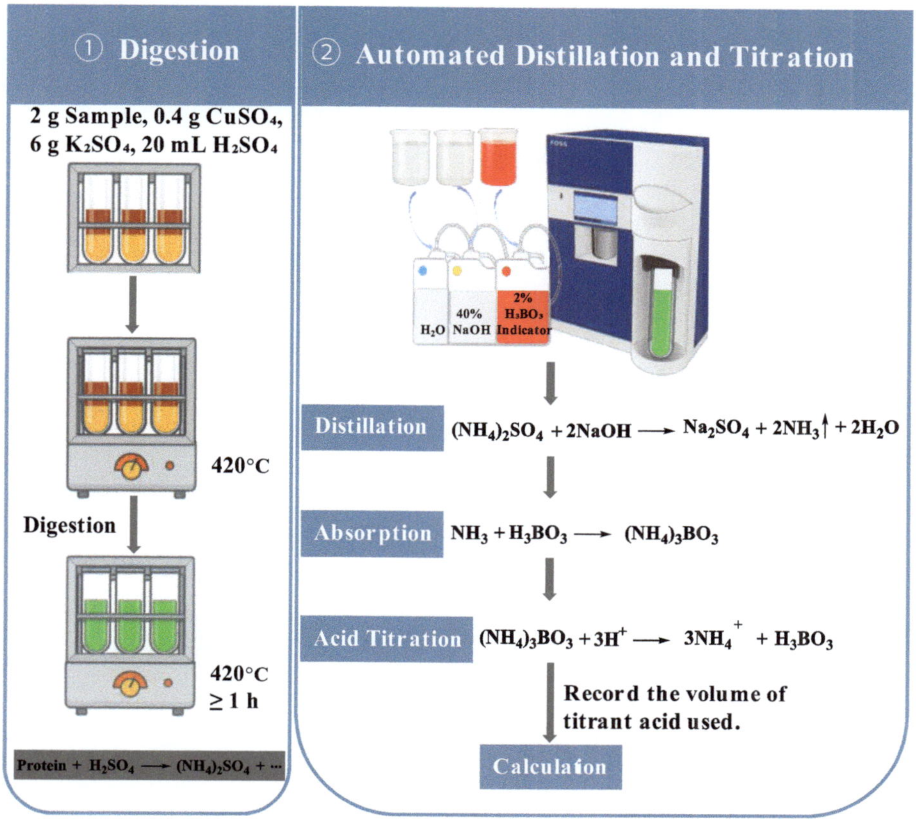

Fig. 4 Schematic diagram of protein content determination by Kjeldahl method

5. Instrument Preparation
    1. Preheat the instrument according to the manufacturer's instructions.
    2. Check the water level in the steam generator and add distilled water if it is low.
    3. Add 2% (w/v) H_3BO_3 indicator solution to the H_3BO_3 bottle.
    4. Add 40% (w/v) NaOH solution to the NaOH bottle.
    5. Fill the titration burette with the standard acid titrant (0.05 mol/L H_2SO_4 or 0.1 mol/L HCl solution).
6. Parameter Settings: optimize parameters according to different instruments. Example parameters for a Foss Kjeltec 8400 analyzer are provided in Table 1.

Testing Procedure

7. Load the cooled digestion tube into the instrument's digestion tube holder and close the door.

Table 1
Foss Kjeltec 8400 parameters

Parameter	Setting
Sample mass (g)	2.0 ± 0.001
Acid titrant	0.05 M H_2SO_4 (preferred)
NaOH volume	80 mL (4× digestion solution volume)
Dilution water	50 mL
Distillation time	≥5 min (ensure complete conversion of $(NH_4)_2SO_4$ to NH_3)
N-to-protein conversion factor	see Table 2 (*see* **Note 21**)

Table 2
N-to-protein conversion factors for cereals in bough and related bread products

Cereals	Ref [27]	Ref [28]	Ref [29]	Ref [30]
Barley	5.83	5.40	5.50	—
Corn	6.25	—	5.65	5.72
Millet, foxtail	—	—	5.80	—
Millet, pearl	—	5.63	5.33	—
Oat	5.83	5.32	5.36	—
Rice	5.95	—	5.17	5.61
Rye	5.83	5.35	5.33	—
Sorghum	—	—	5.65	5.93
Triticale	—	5.62	5.36	
Wheat	5.83	5.49	5.33	5.75
Wheat, bran	6.31	4.71	—	—
Wheat, endosperm	5.70	—	—	—
Wheat, germ	5.80	4.99	—	—

8. Start the program, and the instrument automatically performs:
    1. Rinsing the tube wall with distilled water.
    2. Add 40% NaOH solution to release NH_3.
    3. Steam distill NH_3 into H_3BO_3 absorption solution.
    4. Perform titration with the acid titrant to the endpoint (record consumed volume, V_1).
    5. Calculate %N and %protein.

Blank Test

Reagents other than the test sample are added to the digestion tube as a blank control for testing. The instrument automatically records the volume (V_0) of standard acid titrant consumed by the reagent blank control (*see* **Note 22**).

3.3.3 Dumas Method

The procedure is illustrated in Fig. 5.

Instrument Setup

1. Turn on the instrument and set the following parameters (Table 3):

Instrument Calibration (*see* **Note 23**)

2. Weigh 0.1–0.5 g nitrogen standard substance (aspartic acid or urea, ±0.0001 g).
3. Encapsulate the standard in a tin foil square (seal tightly) using forceps.
4. Place the tin foil packet containing the standard substance on the sample tray of the Dumas instrument and start the "calibration program."
5. The instrument automatically performs the steps of sample feeding, combustion, reduction, separation, detection, and recording the deviation between the theoretical nitrogen content and the measured value of the standard substance (verify deviation ≤1%).
6. Recalibrate if deviation >1%.

Sample Analysis

7. Weigh 0.1–0.5 g (±0.0001 g) of dried sample (*see* **Note 24**).
8. Encapsulate the sample in a tin foil square (seal tightly) using forceps.
9. Place the tin foil packet containing the sample on the sample tray of the Dumas instrument (in the same batch as the calibration sample) and initiate sample analysis.
10. The instrument automatically performs the steps of sample feeding, combustion, reduction, separation, and detection, automatically calculating nitrogen content (%N) (*see* **Note 25**).

3.3.4 Calculation

For the **Kjeldahl method**, protein content (dry basis, %) is calculated using Eq. (3) (*see* **Note 26**):

$$\%\mathbf{Protein} = \%\mathbf{N} \times \mathbf{F} = \frac{(\mathbf{V_1} - \mathbf{V_0}) \times c \times \mathbf{0.014}}{m} \times \mathbf{F} \times \mathbf{100} \quad (3)$$

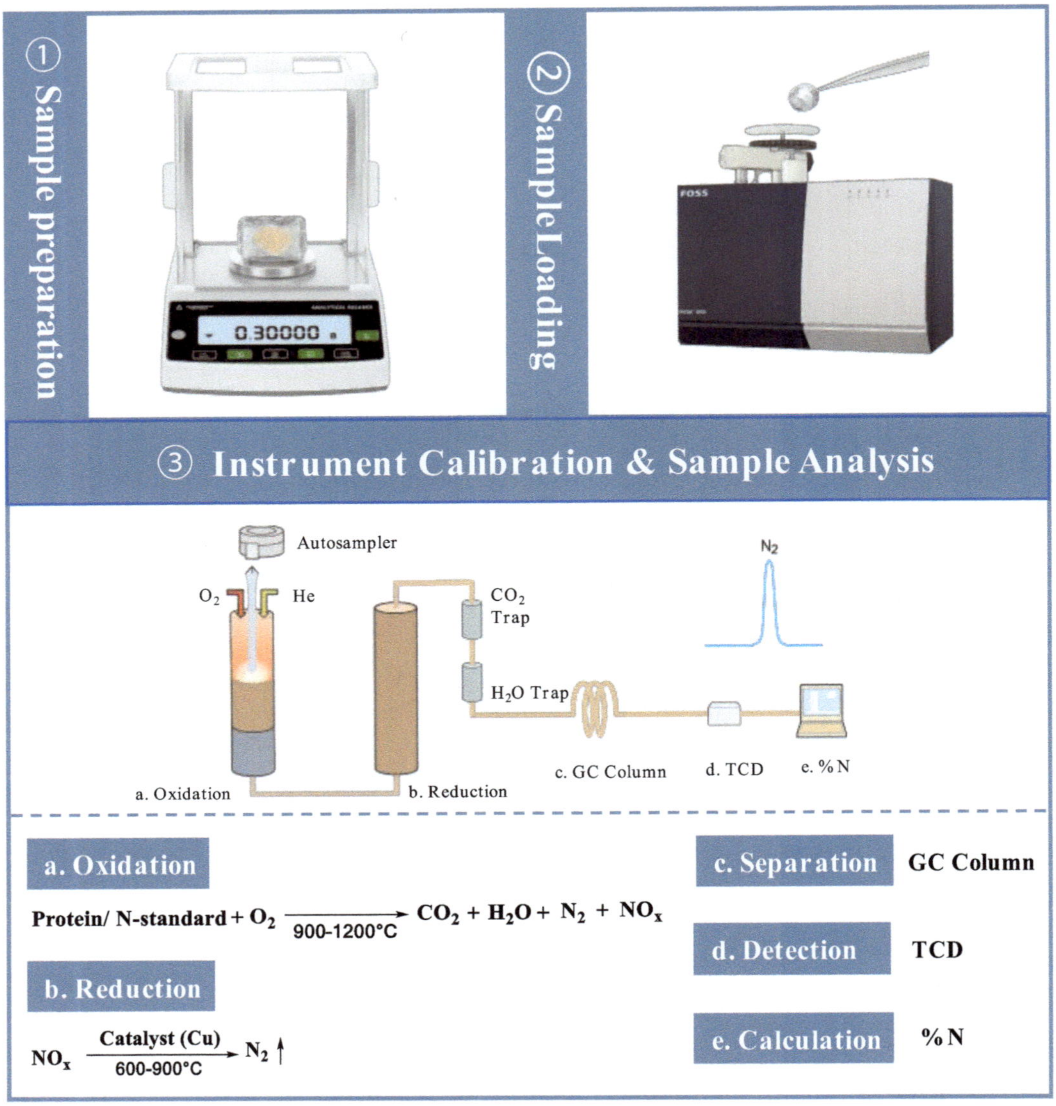

Fig. 5 Schematic diagram of protein content determination by Dumas method

Table 3
Instrument parameter settings for Dumas method

Parameter	Range	Optimization Guidance
Combustion temperature	900–1200°C	Adjust per sample matrix
Reduction temperature	600–900°C	—
O_2 flow rate	Instrument-specific	Target 2%–8% residual O_2 postcombustion
O_2 duration	Sample-dependent	Ensure complete combustion

in which,
%protein is the crude protein content (%).
%N is the nitrogen content (%).
V_1 is the volume of acid titrant (0.05 mol/L H_2SO_4 or 0.1 mol/L HCl solution). consumed by the test solution (mL).
V_0 is the volume of acid titrant (0.05 mol/L H_2SO_4 or 0.1 mol/L HCl solution). consumed by the reagent blank (mL).
c is the concentration of the standard titration solution (mol/L).
0.014 is the molar mass of nitrogen (g/mmol).
m is the mass of the sample (g).
F is the N-to-protein conversion factor
100 is the conversion factor from g/g to g/100 g.

For the **Dumas method**, protein content (dry basis, %) is calculated using Eq. (4):

$$\mathbf{\%Protein = \%N \times F} \quad (4)$$

in which,

1. %protein is the crude protein content (%).
2. %N is the nitrogen content (%).
3. F is the N-to-protein conversion factor.
4. *Reporting*: Express results to 0.01% precision.

3.4 Crude Fat Determination [31, 32]

Soxhlet Extraction Method

The procedure is illustrated in Fig. 6.

3.4.1 Sample Preparation

Homogenize and dry the sample according to Sect. 3.1 (Air-Oven Method).

3.4.2 Extraction Flask Taring

1. Dry a clean Soxhlet extraction flask in an oven at 103 ± 2°C for 2 h.
2. Remove the flask and place it in a desiccator to cool for 30 min, then weigh it.
3. Repeat the drying, cooling, and weighing steps until constant mass. Record the tare mass (m_0).

3.4.3 Sample Loading

1. Weigh 2–5 g of the dried, homogenized sample (record as m) and enclose it in a cartridge made of defatted filter paper, carefully forming it into a cylindrical shape (*see* **Note 27**).
2. Plug the open end of the cartridge with a small amount of defatted cotton to prevent sample loss during extraction.

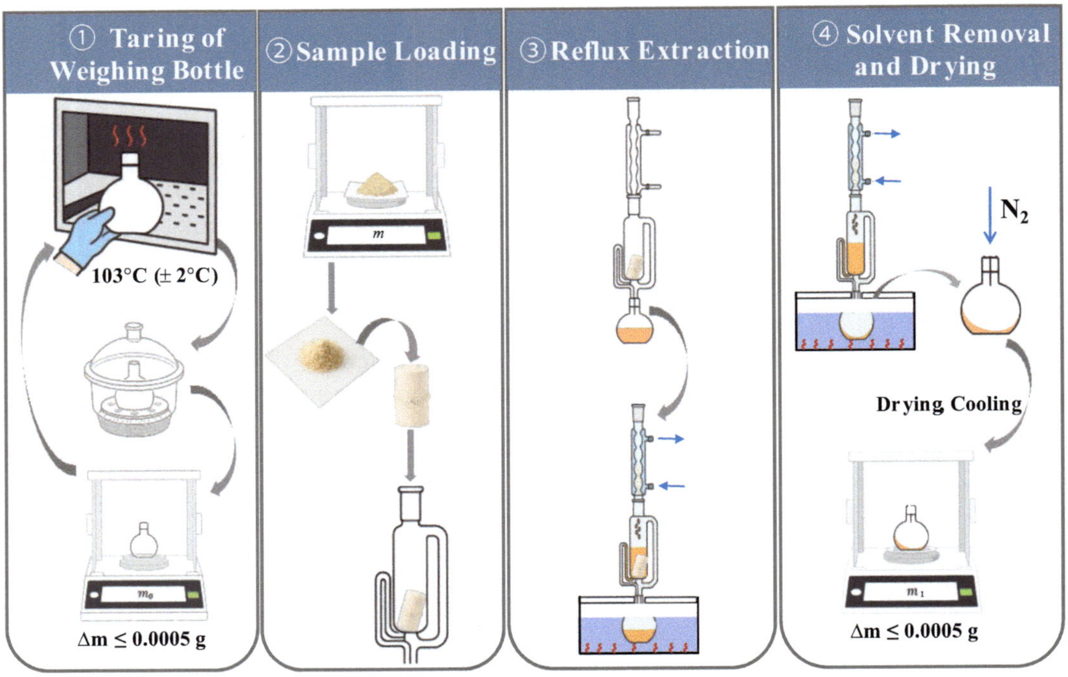

Fig. 6 Schematic diagram of crude fat content determination by Soxhlet extraction method

3. Insert the prepared filter paper cartridge containing the sample into the extraction chamber of the Soxhlet apparatus.

3.4.4 Reflux Extraction

1. Attach the preweighed extraction flask, which has reached a constant weight, as the receiving vessel.
2. Add petroleum ether through the top of the condenser until the solvent in the flask occupies approximately two-thirds of its volume.
3. Assemble the extraction chamber and condenser securely, and start the cooling water flow through the condenser (*see* **Note 28**).
4. Heat the extraction flask in a thermostatically controlled water bath to bring the petroleum ether to gentle boiling. Adjust the heating to achieve 4–6 extraction cycles/h. Continue extraction for 6–8 h (*see* **Note 29**).
5. To verify that extraction is complete, use a ground glass rod to place a drop of the extract onto clean filter paper. The absence of an oily ring on filter paper confirms complete extraction.

3.4.5 Solvent Removal and Drying

1. Upon completion of extraction, stop heating and allow the solvent in the extraction chamber to drain back into the flask. Detach the flask carefully.

2. Evaporate the petroleum ether from the flask at a temperature below its autoignition point. It is advisable to use a nitrogen purge device to accelerate the removal of residual solvents (the nitrogen flow should just disturb the liquid surface).
3. Dry the flask in an oven at 103 ± 2°C for 1 h, then cool to room temperature in a desiccator for 30 min.
4. Repeat the drying, cooling, and weighing cycles until constant mass. Record the mass (m_1) of the flask and fat.

3.4.6 Calculation

Crude fat content (dry basis, %) is calculated using Eq. (5):

$$\%Fat = \frac{m_1 - m_0}{m} \times 100 \quad (5)$$

in which,
%Fat is the crude fat content in the sample (g/100 g dry basis).
m_1 is the mass of extraction flask plus extracted fat (g).
m_0 is the mass of extraction flask prior to extraction (g).
m is the mass of the dry test sample (g).

Moisture Correction (for Wet Samples):

$$m = m_W \times (1 - \%Moisture)$$

m_w is the mass of the wet test sample (g).
Reporting: Express results to 0.1% precision.

3.5 Dietary Fiber Determination [33–35]

3.5.1 Sample Preparation

Enzymatic-Gravimetric Method

The procedure is illustrated in Fig. 7A.

1. Dehydration:
   Dry the sample (≥50 g) and calculate the moisture content as described in Sect. 3.1.
2. Defatting (fat content >10%) (*see* **Note 30**):
   1. Add 50 g (±0.001 g) sample to 1000 mL conical flask with 500 mL petroleum ether (30–60°C fraction).
   2. Shake for 2 min, let stand for 10 min, remove petroleum ether. Repeat defatting twice (three times total).
   3. Dry the defatted sample and calculate the fat content.

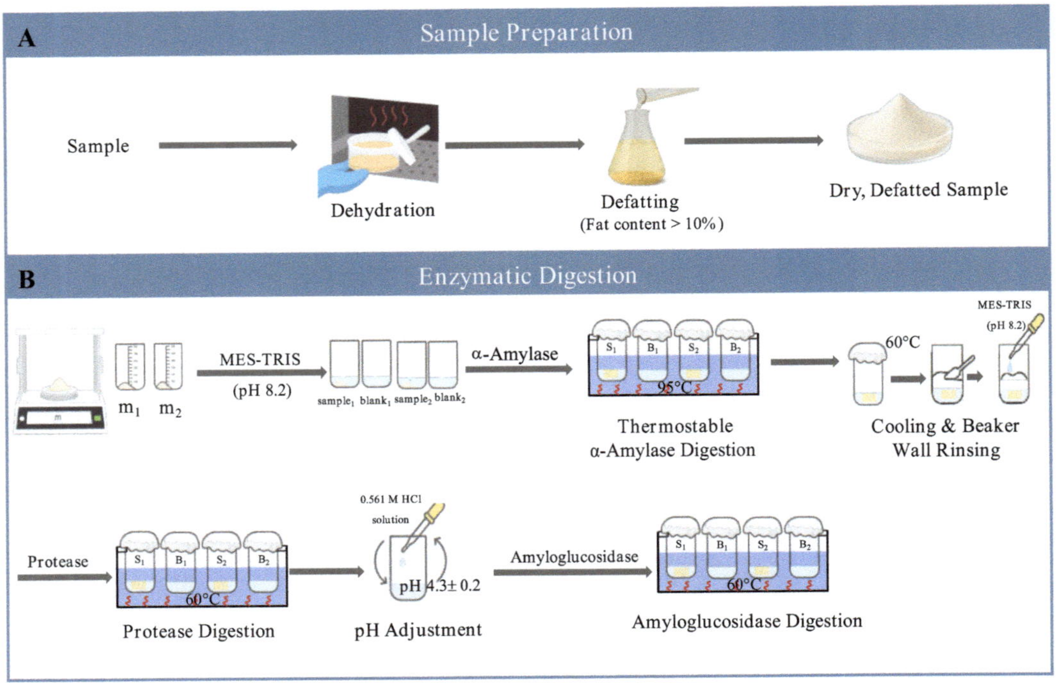

Fig. 7 Schematic illustration of the sample preparation (**A**) and enzymatic digestion (**B**) procedures for dietary fiber determination using the Enzymatic-Gravimetric method

3. Store the dehydration and defatted samples in sealed containers in a desiccator until analysis.
4. Correct final %DF after deducting the %moisture and %fat.

3.5.2 Enzymatic Digestion

The procedure is illustrated in Fig. 7B.

1. Weigh two portions of 1.000 g (±0.001 g) dry and defatted samples into 400–600 mL tall-form beaker.
2. Run two blank controls (reagents only) per assay to account for reagent contributions to the residue.
3. Add 40 mL of MES-TRIS buffer (pH 8.2) to each beaker and stir magnetically until the sample is completely dispersed in the solution (*see* **Note 31**).
4. Thermostable α-Amylase Digestion:
   1. Add 100 μL thermostable α-amylase solution to each beaker with low speed stirring.
5. Cover the beakers with aluminum foil and incubate in a shaking water bath at 95–100°C for 35 min. Begin timing once all beakers have been placed in the water bath at 95°C.

6. Cooling and Beaker Wall Rinsing:
   1. Remove all sample beakers from the water bath and cool to 60°C (*see* **Note 32**).
   2. Open the aluminum foil lid and use a spatula to scrape away the paste from the inside of the beaker, as well as the gel from the bottom of the beaker.
   3. Rinse the beaker walls and the scraper with 10 mL of MES-TRIS buffer (pH 8.2).
7. Protease Digestion:
   1. Add 100 μL protease solution to each beaker.
   2. Recover with aluminum foil and continue shaking in a water bath at 60°C (±1°C) for 30 min. Ensure the water bath has stabilized at 60°C before commencing this step.
8. pH Adjustment:
   1. Stop shaking and open the aluminum foil cover.
   2. Add 5 mL of 0.561 mol/L HCl solution to each beaker while stirring. Adjust the final pH to 4.3±0.2 using 1 mol/L NaOH or HCl solution (*see* **Note 33**).
9. Amyloglucosidase Digestion:
   1. Add 100 μL amyloglucosidase solution while stirring.
   2. Recover with aluminum foil and continue shaking in a water bath at 60°C (±1°C) for 30 min (the water bath temperature should be 60°C at the start).

3.5.3 Total Dietary Fiber (TDF) Determination

The procedure is illustrated in Fig. 8.

1. Precipitation of Dietary Fiber with Ethanol:
   1. Add 225 mL preheated 95% ethanol (60°C) to each of the digestate from Sect. 3.5.2 (4:1 v/v ethanol:digestate).
   2. Remove the beakers and recover with aluminum foil.
   3. Precipitate at room temperature for 1–2 h.
2. Filtration:
   1. Prewet and redistribute celite in a treated and preweighed crucible with 15 mL of 78% ethanol.
   2. Filter the sample ethanol precipitate mixture through the crucible into a collection flask. Transfer all residue from the beaker to the crucible using a spatula and 78% ethanol.

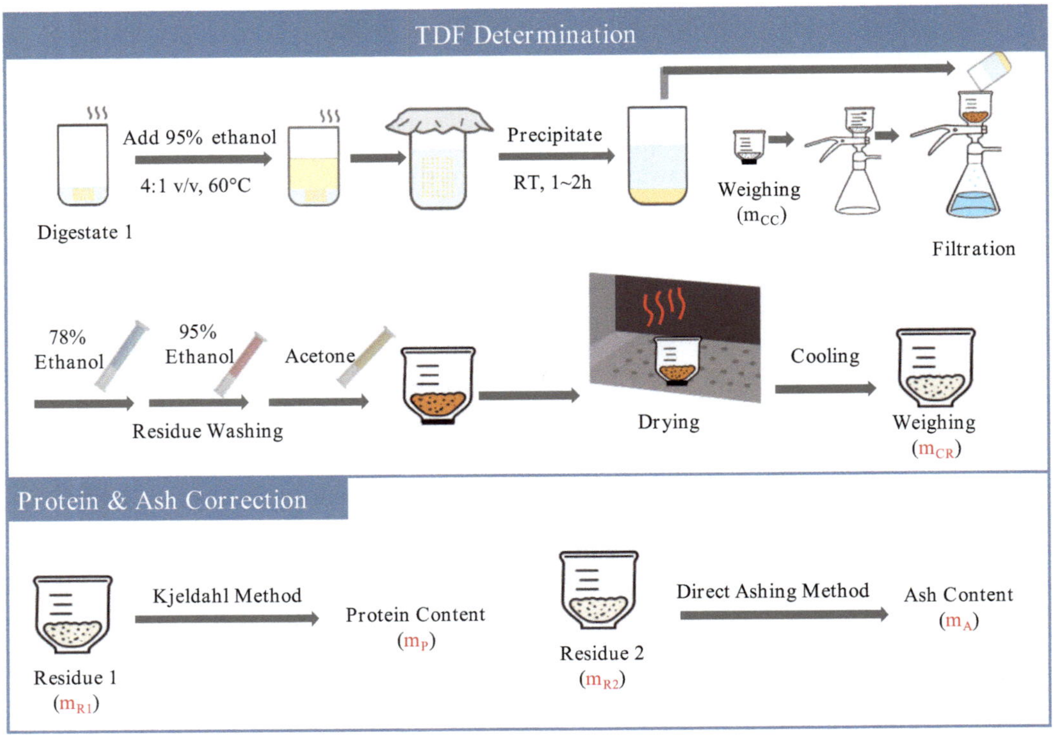

Fig. 8 Schematic illustration of TDF determination using the Enzymatic-Gravimetric method

3. Residue Washing and Drying:
   1. Rinse the residue twice with 15 mL of 78% ethanol, 95% ethanol, and acetone, respectively. Remove the washing solutions by filtration.
   2. Dry the crucible (with residue) overnight in an oven at 103 ± 2°C.
   3. Cool the crucible to room temperature in a desiccator.
   4. Weigh the crucible containing dietary fiber residue and celite (m_{CR}, ±0.1 mg). Calculate the weight of the residue (m_R, ±0.1 mg) by subtracting the weight of the crucible and celite (m_{CC}, ±0.1 mg).
4. Protein and Ash Correction:
   1. One sample residue is taken to determine the protein content (m_P, ±0.1 mg) using Kjeldahl method described in Sect. 3.3.2. The N-to-protein conversion factor is 6.25 in all cases.
   2. Another parallel sample residue is taken to determine the ash content (m_A, ±0.1 mg) using the dry ashing method described in Sect. 3.2.

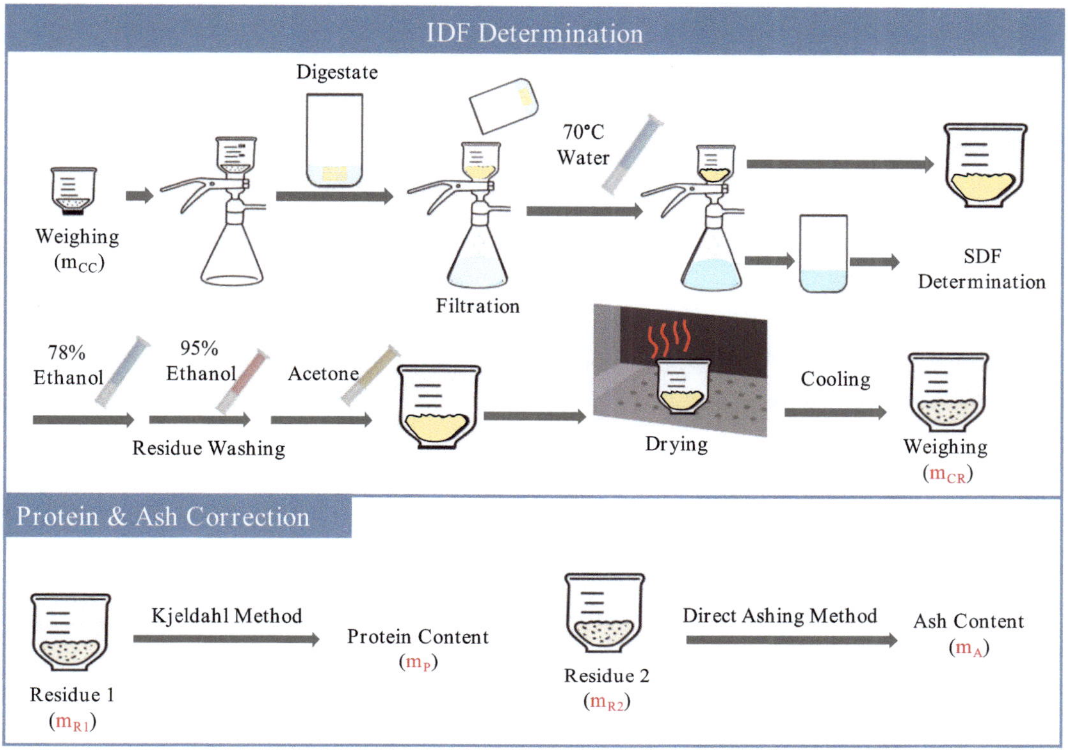

Fig. 9 Schematic illustration of IDF determination using the Enzymatic-Gravimetric method

3.5.4 Insoluble Dietary Fiber (IDF) Determination

The procedure is illustrated in Fig. 9.

1. Filtration:
    1. Prewet and redistribute celite in a treated and preweighed crucible with 3 mL of water.
    2. Connect the crucible to a filtration apparatus and draw off the water to ensure the celite is evenly spread on the filter plate.
    3. Filter the digestate from Sect. 3.5.2 through the crucible into a collection flask. Wash the residue twice with 10 mL distilled water preheated to 70°C.
    4. Transfer the filtrate and wash solution to a 600 mL tall beaker for soluble dietary fiber determination (follow the method described in Sect. 3.5.5).
2. Residue Washing and Drying:
   Proceed as described in the Step 3 in Sect. 3.5.3 (*see* **Note 34**).
3. Protein and Ash Correction:

Determination according to the Step 4 in Sect. 3.5.3.

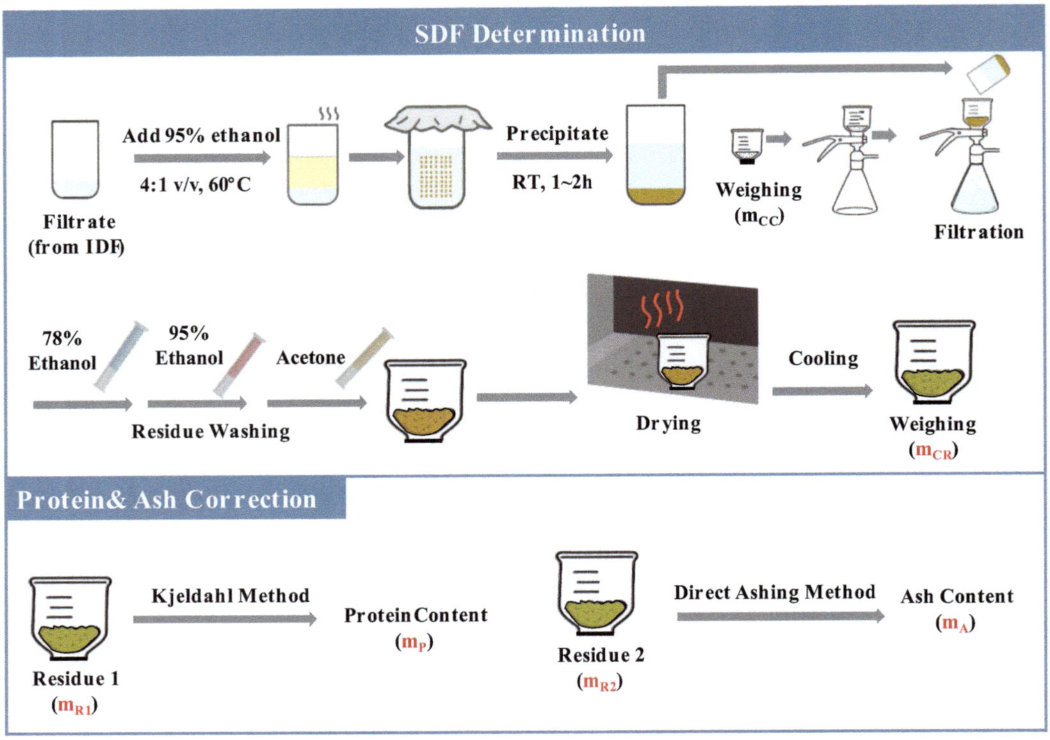

Fig. 10 Schematic illustration of SDF determination using the Enzymatic-Gravimetric method

3.5.5 Soluble Dietary Fiber (SDF) Determination

The procedure is illustrated in Fig. 10.

1. Filtrate Handling:
   1. Weigh the mixture of filtrate and wash solution collected in the IDF determination procedure.
   2. Calculate the mass of the mixed solution by deducting the mass of the beaker and estimate its volume (*see* **Note 35**).
2. Precipitation:
   1. Add about four times the filtrate volume of preheated (60°C) 95% ethanol.
   2. Allow precipitation to form at room temperature for more than 1 h.
3. Filtration, Washing, and Drying:
   Follow Steps 2–4 of the TDF determination procedure.
4. Protein and Ash Correction:

Subsequent operations were performed according to Steps 2–4 of the TDF determination (Sect. 3.5.3).

3.5.6 Calculation

TDF, IDF, and SDF contents (dry basis, %) are using Eq. (6):

$$\text{Dietary fiber}(\%) = \frac{\frac{m_{R1} + m_{R2}}{2} - m_P - m_A - m_B}{\frac{m_1 + m_2}{2}} \times 100 \quad (6)$$

where m_{R1} = residue weight 1 from sample 1 = m_{CR1}–m_{CC1}, m_{R2} = residue weight 2 from sample 2 = m_{CR2}–m_{CC2}, m_1 = sample weight 1, m_2= sample weight 2, m_A = ash weight from residue 1, m_P = protein weight from residue 2, and

$$m_B = \frac{m_{BR1} + m_{BR2}}{2} - m_{BP} - m_{BA}$$

where m_{BR1} = blank residue 1, m_{BR2} = blank residue 2, m_{BP} = blank protein from blank residue 1, and m_{BA} = blank ash from blank residue 2.

Reporting: Express results to 0.01% precision.

4 Notes

1. Avoid using ceramic crucibles as they can absorb moisture, leading to inaccurate results.
2. Silica gel and anhydrous $CaCl_2$ are not suitable desiccants.
3. The ratio of methyl red to methylene blue or bromocresol green may be adjusted based on specific apparatus requirements.
4. H_2SO_4 (0.05 mol/L) is generally preferred over HCl for titration in Kjeldahl systems, as it minimizes bubble formation in delivery tubing.
5. Select size based on sample mass, 150 mL for 2–3 g samples, 250 mL for 3–5 g samples.
6. Petroleum ether is highly flammable. Take extreme care not to allow vapor concentrations to reach ignition levels in the work area (explosion risk from static sparks).
7. The pH of MES-TRIS buffer changes with temperature. Buffer pH decreases from 8.2 (24°C) to 7.4–7.6 (55–60°C) and 6.9–7.2 (85–90°C). This gradient aligns with the optimal pH of thermostable α-amylase (pH 6.0 at 60°C → pH 7.0 at 90°C), eliminating pH adjustments during protease incubation.

8. Perform sampling and preparation rapidly (total time ≤ 30 min) to prevent moisture absorption, especially in high-humidity environments.
9. Release vacuum slowly to prevent sample splashing.
10. For high-sugar samples, which are prone to moisture absorption, cooling, and weighing must be performed quickly.
11. If parallel sample results vary by more than 5%, resample and reprocess.
12. Required if reporting results on dry basis.
13. Control heating rate to prevent foaming or sample loss. Excessive bubbling elevates ash yield.
14. The ashing temperature should not exceed 575°C (as high temperatures may cause loss of volatile minerals such as potassium and sodium, yielding lower results).
15. For the Dumas method, samples must be dried to constant weight (moisture ≤10%). Moisture evaporation absorbs combustion heat (reducing temperature and causing incomplete organic decomposition) and reacts with NO_x to form HNO_3 (impairing nitrogen reduction, leading to low results).
16. Perform in acid-resistant fume hood with PPE (gloves, goggles, lab coat).
17. (a) Adjust the sample weight to minimize titration error if the protein content is too low and the titration volume is less than 1 mL. (b) Adjust the sample weight to prevent or minimize boiling, splashing, or spilling during digestion if the sample has a high sugar content.
18. (a) Avoid splashing by adding H_2SO_4 along the tube wall. (b) Concentrated H_2SO_4 is highly corrosive, gloves, and safety goggles should be worn during operation. (c) The acid volume can be adjusted according to the equipment optimization parameters.
19. Add another 2 mL of concentrated H_2SO_4 and extend the digestion time if the digestive solution does not clear.
20. For rapid cooling, place the bottom of the digestion tube under running water.
21. If the conversion factor is unknown, use 6.25 (default). For cereals, a smaller factor (e.g., 5.60) may be more appropriate.
22. 1 g of sucrose can be used instead of the test sample in the blank control.
23. (a) The detection sensitivity of the Dumas instrument decreases with use time, requiring regular calibration with nitrogen standards (every 10–20 samples). (b) The weight of

the standard should match the weight of the sample to ensure the accuracy of the calibration curve.

24. Adjust based on protein content:
    (a) High-protein (>12%): 0.1–0.2 g.
    (b) Low-protein (<8%): 0.3–0.5 g.
25. If the sample contains phosphorus or sulfur compounds, SO_2 and P_2O_5 will be generated after combustion, interfering with TCD detection (resulting in large nitrogen content deviation). The SO_2 can be removed using a sulfur adsorbent (e.g., sodium hydroxide pellets, some Dumas instruments have a built-in sulfur adsorbent).
26. (a) Express the result to two decimal places. (b) State F value when reporting %protein. (c) The moisture data (Sect. 3.1) is required to correct the protein content (on a dry basis, %) of undried samples.
27. Ensure the cartridge height is lower than the siphon arm of the Soxhlet apparatus to prevent overflow.
28. Ensure a continuous flow of cooling water to the condenser to prevent solvent loss through evaporation.
29. The recommended frequency is 4–6 cycles/h. Extraction that is too rapid may result in incomplete sample soaking, while extraction that is too slow may unnecessarily prolong the procedure.
30. The fat content in bread is usually 5%–15%, and 3%–8% in frozen dough. It is necessary to judge whether to defat or not according to the actual situation.
31. Stir thoroughly to prevent sample clumping to ensure enzyme accessibility.
32. Adjust the water bath temperature to 60°C by draining some hot water and adding cold water.
33. Measure and adjust pH at 60 ± 1°C, as the pH will be higher at lower temperatures.
34. Wash the residue with 78% ethanol, 95% ethanol, and acetone promptly to prevent an increase in the value of insoluble dietary fiber.
35. Preweigh the 600 mL beaker before collecting the filtrate.

Acknowledgments

This work was supported by the Science and Technology R&D Program Joint Fund Project of Henan Province, China (No. 232103810062); the High-level Talent Start-up Project of Henan

University of Technology, China (No. 2023BS102); the Postdoctoral Research Project of Henan Province, China (No. HN2025051); the Agriculture Research System of China of MOF and MARA (No. CARS-14-1-29); and the Cultivation Project of Tuoxin Team in Henan University of Technology (No. 2024TXTD03).

References

1. Rajput I, Bharti A, Patel R, Parihar P, Shukla SS et al (2024) Combined effects of hydrocolloids, frozen storage and thawing on proximate composition of frozen dough bread using response surface methodology. Int J Adv Biochem Res 8:233–240
2. Jiang J, Yang J, Fan X, Zhang Y, Li M, Zhang B et al (2025) Regulation of ice crystal growth in frozen dough: from the effect of gluten and starch fractions interaction on water binding—a review. Food Chem 476:143509
3. Schiraldi A, Fessas D (2012) 14—The role of water in dough formation and bread quality. In: Cauvain SP (ed) Breadmaking. Second. Woodhead Publishing, pp 352–369
4. Liu H, Xing Y, Liu D, Yang Q, Xiao S, Fu Y et al (2023) Quality deterioration of frozen dough bread during terminal freezing and thawing: from the insight of moisture and starch properties. Food Biosci 56:103435
5. AACC Approved Methods of Analysis, 11th edn. Method 44–15.02. Moisture—Air-Oven Methods. Approved 2010. Cereals & Grains Association, St. Paul, MN, USA
6. ISO (2024) ISO 712-1:2024 Cereals and cereal products—Determination of moisture content. International Organization for Standardization, Geneva, Switzerland
7. Cardoso RVC, Fernandes Â, Heleno SA, Rodrigues P, Gonzaléz-Paramás AM, Barros L et al (2019) Physicochemical characterization and microbiology of wheat and rye flours. Food Chem 280:123–129
8. Li J, Zhang S, Liu C, Yin Y, Sun X, Wu J (2023) Characterization of ash content in wheat flour using data fusion. Infrared Phys Techn 133:104792
9. Clément H, Prost C, Chiron H, Ducasse MB, Della Valle G, Courcoux P et al (2018) The effect of organic wheat flour by-products on sourdough performances assessed by a multi-criteria approach. Food Res Int 106:974–981
10. Marshall MR (2010) Ash analysis. In: Nielsen SS (ed) Food Analysis. Springer US, Boston, MA, pp 105–115
11. Ribotta PD, León AE, Añón MC (2001) Effect of freezing and frozen storage of doughs on bread quality. J Agr Food Chem 49:913–918
12. Beljkaš B, Matić J, Milovanović I, Jovanov P, Mišan A, Šarić L (2010) Rapid method for determination of protein content in cereals and oilseeds: validation, measurement uncertainty and comparison with the Kjeldahl method. Accredit Qual Assur 15:555–561
13. Punzalan JM, Hartono P, Fraser-Miller SJ, Leong SY, Sutton K, Moggre G et al (2025) MicroNIR spectroscopy with chemometric analysis provides rapid protein content evaluation and prediction of semi-refined flaxseed protein extract produced via pulsed electric field (PEF)-assisted extraction. Food Chem 473:143062
14. Serrano S, Rincón F, García-Olmo J (2013) Cereal protein analysis via Dumas method: standardization of a micro-method using the EuroVector elemental analyser. J Cereal Sci 58:31–36
15. Pareyt B, Finnie SM, Putseys JA, Delcour JA (2011) Lipids in bread making: sources, interactions, and impact on bread quality. J Cereal Sci 54:266–279
16. Mamat H, Hill SE (2014) Effect of fat types on the structural and textural properties of dough and semi-sweet biscuit. J Food Sci Tech Mys 51:1998–2005
17. Jiang Y, Zhao Y, Zhu Y, Qin S, Deng Y, Zhao Y (2019) Effect of dietary fiber-rich fractions on texture, thermal, water distribution, and gluten properties of frozen dough during storage. Food Chem 297:124902
18. Wu C, Wu F, Zhang Y, Liu F, Luan G (2025) Effect of extruded soybean okara on the texture, rheology, and structural properties of high-fiber composite dough. Food Chem 487:144722
19. National Health Commission of the People's Republic of China, State Administration for Market Regulation of the People's Republic of China (2016) GB 5009.3-2016 National Food Safety Standard—determination of Moisture in Foods (Method I: Direct drying method; Method II: Vacuum drying method). China Standards Press, Beijing, China
20. AACC Approved Methods of Analysis, 11th edn. Method 08–01.01. Ash—Basic Method. Approved 2010. Cereals & Grains Association, St. Paul, MN, USA

21. ISO (2023) ISO 2171:2023(E). Cereals, pulses and by-products—determination of ash yield by incineration. International Organization for Standardization, Geneva, Switzerland
22. National Health Commission of the People's Republic of China, State Administration for Market Regulation of the People's Republic of China. (2016). GB 5009.4-2016 National Food Safety Standard—Determination of Ash in Food (Method I: Determination of total ash in food). China Standards Press, Beijing, China
23. AACC Approved Methods of Analysis, 11th edn. Method 46–11.02.Crude Protein—Improved Kjeldahl Method, Copper Catalyst Modification. Approved 2010. Cereals & Grains Association, St. Paul, MN, USA
24. AACC Approved Methods of Analysis, 11th edn. Method 46–30.01. Crude Protein—Combustion Method. Approved 2010. Cereals & Grains Association, St. Paul, MN, USA
25. ISO (2013). ISO 20483:2013(E). Cereals and pulses-Determinationof the nitrogen content and calculationof the crude protein content-Kjeldahl method. International Organization for Standardization, Geneva, Switzerland
26. National Health Commission of the People's Republic of China, State Administration for Market Regulation of the People's Republic of China. (2025). GB 5009.5-2025 National Food Safety Standard—Determination of Protein in Food (Method I: Kjeldahl method; Method III: Dumas combustion method). China Standards Press, Beijing, China
27. Jones DB (1931) Factors for Converting Percentages of Nitrogen in Foods and Feeds Into Percentages of Proteins. USDA Circ 183(1):1–21
28. Tkachuk R (1969) Nitrogen-to-protein conversion factors for cereals and oilseed meals. Cereal Chem 46:419–424
29. Mosse J (1990) Nitrogen-to-protein conversion factor for ten cereals and six legumes or oilseeds. A reappraisal of its definition and determination. Variation according to species and to seed protein content. J Agr Food Chem 38:18–24
30. Sosulski FW, Imafidon GI (1990) Amino acid composition and nitrogen-to-protein conversion factors for animal and plant foods. J Agr Food Chem 38:1351–1356
31. AACC Approved Methods of Analysis, 11th edn. Method 30–25.01. Crude fat in wheat, corn, and soy flour, feeds, and mixed feeds. Approved 2010. Cereals & Grains Association, St. Paul, MN, USA
32. National Health Commission of the People's Republic of China, State Administration for Market Regulation of the People's Republic of China. (2016). GB 5009.6-2016 National Food Safety Standard—Determination of fat in food (Method I: Soxhlet Extraction Method). China Standards Press, Beijing, China
33. AACC Approved Methods of Analysis, 11th edn. Method 32–07.01. Soluble, insoluble, and total dietary fiber in foods and food products. Approved July 2010. Cereals & Grains Association, St. Paul, MN, USA
34. AOAC Official Methods, 19th edn. Method 2011.25. Insoluble, soluble, and total dietary fiber in foods: enzymatic-gravimetric-liquid chromatography. Approved 2011. AOAC International, Gaithersburg, MD, USA
35. National Health Commission of the People's Republic of China, State Administration for Market Regulation of the People's Republic of China. (2023). GB 5009.88-2023 National Food Safety Standard—Determination of Dietary Fiber in Food (Method I: Enzymatic-Gravimetric Method). China Standards Press, Beijing, China

Check for updates

Chapter 10

Free Amino Acids Analysis

Feiyu An, Rina Wu, Junrui Wu, Yuxuan Liu and Shuqi An

Abstract

Free amino acids (FAAs) are core indicators regulating the flavor quality, nutritional value, and shelf-life stability of bread. Their content and compositional characteristics can directly characterize the scientificity and rationality of bread processing techniques such as fermentation and baking. Therefore, establishing accurate analytical methods for FAAs in bread is of great significance. High performance liquid chromatography (HPLC), with its significant advantages of high sensitivity, high analytical efficiency, and wide applicability, has become one of the most commonly used and effective techniques for amino acid analysis. However, FAAs in bread lack inherent ultraviolet (UV) or fluorescent chromophores, leading to low sensitivity when detected directly by HPLC, making it difficult to meet the demands of precise analysis. Thus, derivatization pretreatment becomes a key step in the detection of such substances. This chapter systematically elaborates an analytical method based on the combination of OPA/FMOC precolumn derivatization technology and HPLC, aiming to provide a technical reference for the efficient and accurate quantification of FAAs in the bread matrix, and to lay a methodological foundation for subsequent research on bread processing optimization and quality control.

Keywords Free amino acids, High performance liquid chromatography, Bread, Precolumn derivatization

1 Introduction

Bread, as a widely consumed food, has its nutritional quality and flavor characteristics influenced by various factors, among which the composition and content of FAAs are important indicators [1]. FAAs not only directly affect the flavor and color of bread but also reflect the degree of protein breakdown during processing and its nutritional value. For example, glutamic acid and aspartic acid impart umami taste to bread, while essential amino acids, such as leucine and isoleucine are crucial for human nutritional needs. Therefore, accurately determining the content of FAAs in bread is highly significant for controlling product quality, optimizing formulations, assessing nutritional value, and studying flavor formation mechanisms.

Guoqiang Zhang and Jinying Guo (eds.), *Frozen Dough and Bread Quality*, Methods and Protocols in Food Science,
https://doi.org/10.1007/978-1-0716-5206-0_10,

Currently, FAA detection has formed a multidimensional technical system, mainly divided into two categories: one is traditional chemical analysis methods, including spectrophotometry (such as the ninhydrin colorimetric method) and formaldehyde titration. Although these methods are simple to operate, they have limitations in specificity, sensitivity, and the ability to simultaneously detect multiple components. The other category is modern instrumental analysis techniques, including amino acid analyzer methods, capillary electrophoresis, high-performance liquid chromatography (HPLC), gas chromatography (GC), and gas chromatography-mass spectrometry (GC-MS) [2]. These methods have become the mainstream application due to their higher separation efficiency, quantitative accuracy, and anti-interference capability. Among them, HPLC is particularly widely used due to its good applicability, reproducibility, and ability to simultaneously analyze multiple components in complex matrices, making it one of the most common and effective techniques for amino acid analysis currently [3]. This chapter will elaborate in detail on the principles, procedures, and key technical points of HPLC for determining FAAs. Since most FAAs lack UV or fluorescent chromophores, leading to low direct detection sensitivity, derivatization processing is usually necessary. Common derivatization methods include precolumn and postcolumn derivatization. Precolumn derivatization is a key step in this experiment, aiming to allow amino acids, which themselves have no fluorescent signal, to react with derivatizing reagents to generate substances with strong fluorescent signals for high-sensitivity detection using a fluorescence detector, suitable for complex food matrices like bread [4, 10]. The combined use of o-phthalaldehyde (OPA) and 9-fluorenylmethyl chloroformate (FMOC) is a common precolumn derivatization strategy. OPA can react with primary amine amino acids (such as aspartic acid, glutamic acid, leucine, etc.), while FMOC can react with secondary amine amino acids (such as proline, hydroxyproline), thus, achieving complete detection of all amino acids and providing an effective analytical tool for the quality analysis and control of bread.

2 Materials

2.1 Instruments and Materials

1. Bread (commercially available breads of different brands)
2. High-performance liquid chromatography system (Agilent, 1290 Infinity II)
3. Chromatographic column (C18, 4.6×100 mm)
4. Ultrasonic extraction
5. o-Phthalaldehyde (OPA)
6. 9-Fluorenylmethyl chloroformate (FMOC)
7. Boric acid

8. Sodium acetate
9. Sodium hydroxide
10. Acetonitrile (HPLC grade)
11. Methanol (HPLC grade)
12. Ultrapure water
13. Analytical balance
14. Syringe filter (0.22 μm)

2.2 Solution Preparation

1. Amino acid standard stock solutions: Precisely weigh each amino acid standard separately, dissolve and dilute to volume with ultrapure water to prepare single standard stock solutions with a concentration of 1.0 mg/mL. Store at 4°C.
2. Mixed amino acid standard working solutions: Before use, take an appropriate amount of each single standard stock solution and dilute with ultrapure water to a series of mixed standard working solutions at different concentrations (*see* **Note 1**).
3. OPA derivatization reagent: Weigh 10 mg OPA and dissolve in 0.5 mL methanol, add 2 mL of 0.4 mol/L sodium borate buffer (pH 9.5), and mix well. Prepare immediately before use (*see* **Notes 2 and 3**).
4. FMOC derivatization reagent: Accurately weigh 5 mg FMOC, dissolve and dilute to volume with acetonitrile to 10 mL, to prepare a 2.5 mmol/L FMOC solution (*see* **Notes 2 and 3**).

3 Methods

3.1 Sample Collection

Place the bread sample to be tested in a dry, clean sealed bag to avoid moisture absorption or contamination [5]. After bringing it back to the laboratory, remove the crust as soon as possible, crush and homogenize the edible part, and set aside for analysis.

3.2 Sample Pretreatment

1. Weigh 1.00 g of uniformly crushed bread into a 50 mL of centrifuge tube.
2. Add 20 mL of ultrapure water and vortex mix for 1 min.
3. Place sample in an ultrasonic water bath for extraction at 50°C for 10 min.
4. After extraction and cooling to room temperature, centrifuge at 10,000 r/min for 10 min and collect the supernatant.
5. Repeat the extraction of the residue with 10 mL ultrapure water once, combine the two supernatants, and dilute to 50 mL with ultrapure water.
6. Filter an appropriate amount of the extract through a 0.22-μm microporous membrane filter for derivatization use (*see* **Note 4**).

3.3 Precolumn Derivatization

Take 10 μL of the filtered sample extract or standard working solution into an insert vial. Sequentially add 20 μL of OPA derivatization reagent, vortex mix for 10 s; let stand for 1 min for reaction, then add 20 μ L of FMOC derivatization reagent, vortex mix for 10 s; let stand for 1 min for reaction, then immediately add 50 μL of 0.1 mol/L sodium acetate solution (pH 4.0) to terminate the reaction (*see* **Note 5**). After vortex mixing, take 1 μL for HPLC analysis [6].

3.4 Chromatographic Conditions

Chromatographic column: C18 reversed-phase column (4.6 × 100 mm, 3 μm)

Mobile phase: Phase A: 20 mM Na_2HPO_4 (pH 7.8)

Phase B: Acetonitrile:Methanol:Ultrapure water = 45:45:10 (v/v/v)

Gradient elution program as listed below:

Time (min)	Phase A (%)	Phase B (%)
0.00	97	3
1.26	97	3
10.7	75.6	24.4
20.0	43	57.0
21.0	0	100
24.0	0	100
24.8	97	3
25.0	97	3

Flow rate: 1.5 mL/min
Column temperature: 45°C
Detector: Fluorescence detector (excitation wavelength: 340 nm, emission wavelength: 450 nm)
Injection volume: 1 μL

The operation after the chromatographic experiment is completed is described in **Note 6**.

3.5 Data Analysis

Amino acids were identified qualitatively by comparing the retention times of components in the sample with those of the standards. Using the external standard method, calibration curves (peak area vs. concentration) for each amino acid were plotted by analyzing a series of concentrations of mixed amino acid standard working solutions. The content of various free amino acids in the bread sample was calculated accordingly.

3.6 Method Accuracy and Detection Limit

3.6.1 Chromatographic Analysis Effect

Under the above chromatographic conditions, 19 amino acid mixed standards achieved baseline separation within 25 min. A representative chromatogram is shown in Fig. 1 (chromatogram of 19 amino acid mixed standards). The retention time, resolution (Rs), and theoretical plate number (N) for each amino acid are listed in Table 1 (chromatographic parameters table for 19 amino acids). As shown in Table 1, the chromatographic peaks were symmetrical and sharp. Except for proline, the resolution of all other amino acids was greater than 1.5, and the theoretical plate numbers were all higher than 30,000, indicating high separation efficiency, excellent column performance, effective differentiation of different types of free amino acids, and meeting the separation requirements for complex free amino acid systems [7].

Figure 2 (overlaid chromatograms of six consecutive injections) shows the overlaid chromatograms of six consecutive injections of amino acid standards. The retention time deviation of the corresponding chromatographic peaks was less than 0.5%, and the relative standard deviation (RSD) of the response values was below 2.0%, with high peak overlap. This proves that the method performs excellently in terms of retention time stability and response value repeatability, reducing detection errors caused by instrument fluctuations, and ensuring the reliability of quantitative results.

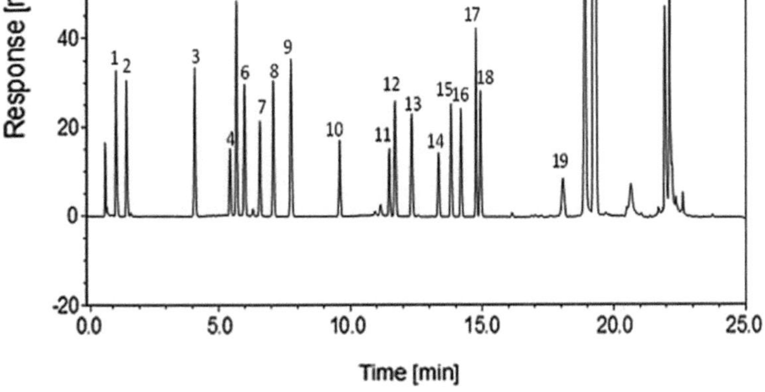

Fig. 1 Chromatogram of 19 amino acid mixed standards

Table 1
Chromatographic parameters table for 19 amino acids

Peak no.	Peak name	Ret. time min	Resol. USP	Plates USP
1	ASP	1.118	5.28	3977
2	GLU	1.508	30.62	6388
3	SER	4.097	14.87	31,594
4	HIS	5.432	2.68	62,977
5	GLY	5.675	3.31	58,291
6	THR	5.991	3.55	62,516
7	ARG	6.578	5.57	91,919
8	TAU	7.084	6.61	90,900
9	ALA	7.753	17.71	84,010
10	TYR	9.596	12.02	148,213
11	CYS	11.497	2.13	281,164
12	VAL	11.708	5.78	181,011
13	MET	12.336	10.08	213,944
14	TRP	13.356	4.93	316,037
15	PHE	13.826	3.94	339,718
16	ILE	14.204	6.62	348,910
17	LEU	14.778	1.88	595,951
18	LYS	14.941	12.62	384,346
19	PRO	18.075	NA	166,358

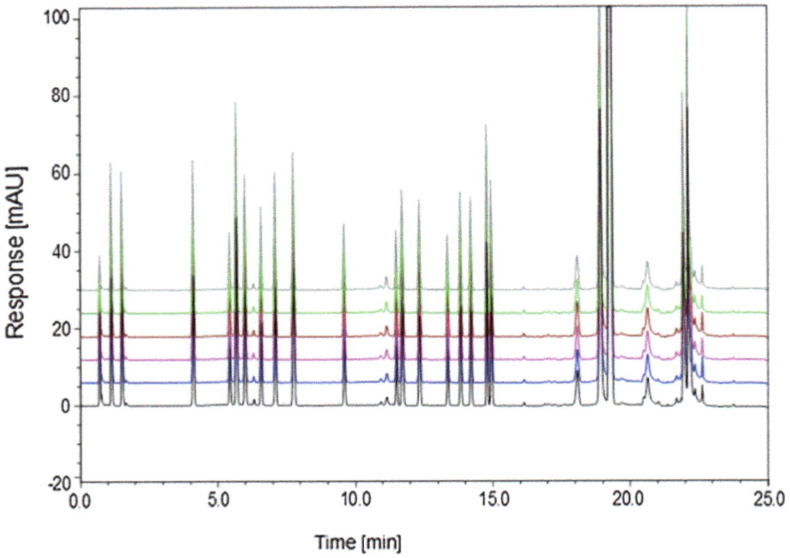

Fig. 2 Overlaid chromatograms of six consecutive injections

3.6.2 Method Linearity and Detection Limit

Standard curves were plotted with the mass concentration (0.5–50 μmol/L) of the 19 amino acid standards as the abscissa and the corresponding peak area as the ordinate, followed by linear regression analysis. The linear correlation coefficients (R^2) for all amino acids were greater than 0.999, indicating a good linear relationship between the mass concentration and peak area within this concentration range, suitable for the quantitative analysis of free amino acids in samples [8].

The method detection limit (LOD) was calculated based on a signal-to-noise ratio of 3 (S/N=3). The LOD range for the 19 amino acids was 0.05–0.20 μmol/L. The method quantification limit (LOQ) was calculated based on a signal-to-noise ratio of 10 (S/N=10). The LOQ range was 0.15–0.60 μmol/L. These detection and quantification limits are far below the conventional content range of free amino acids in actual samples, indicating high method sensitivity and the ability to accurately detect low levels of free amino acids, avoiding missed detection or inaccurate quantification.

3.7 Derivatization Condition Optimization and Verification

The effects of different derivatization conditions (OPA reagent standing time, FMOC reaction time, termination reaction pH) on the detection results were compared. When the OPA reagent was used more than 2 h after preparation, the peak areas of primary amine amino acids decreased by 8.5%–12.3%, indicating that the OPA reagent needs to be prepared fresh before use. When the reaction time of FMOC with secondary amine amino acids was controlled at 1.5 min, the peak area reached its maximum value. The peak area decreased to varying degrees when the reaction time was insufficient or too long, verifying the necessity of strict control of the reaction time. When the termination reaction pH was 4.0, the stability of each amino acid derivative was optimal, and the peak area RSD was the smallest. When the pH deviated from this range, peak area fluctuations increased. Therefore, this pH condition was determined to be the optimal termination reaction parameter [9].

4 Notes

1. Store mixed amino acid standards at −20°C, avoid repeated freeze-thaw cycles.
2. OPA and FMOC are irritating to the skin and mucous membranes. Operate with gloves and goggles in a fume hood.
3. OPA reagent should be prepared fresh before use; FMOC reagent should be protected from light.
4. Starch in bread may clog the filter membrane. After extraction, ensure sufficient centrifugation (≥8000 × *g*). If necessary, use a solid-phase extraction column (e.g., C18 column) for further purification.

5. Derivatization reaction times need strict control. Exceeding the time may lead to derivative decomposition.
6. After the experiment, flush the column with 100% acetonitrile for 30 min to maintain column efficiency. If peak tailing occurs, adding 0.1% triethylamine to mobile phase A may improve it.

References

1. Blandino A, Al-Aseeri M, Pandiella S et al (2003) Cereal-based fermented foods and beverages [J]. Food Res Int 36(6):527–543
2. Azevedo SM, Seraglio TKS, Rocha G et al (2017) Free amino acid determination by GC-MS combined with a chemometric approach for geographical classification of bracatinga honeydew honey (*Mimosa scabrella* Bentham) [J]. Food Control 78:383–392
3. Megías C, Cortés-Giraldo I, Girón-Calle J et al (2014) Determination of β-cyano-L-alanine, γ-glutamyl-β-cyano-L-alanine, and common free amino acids in vicia sativa (fabaceae) seeds by reversed-phase high-performance liquid chromatography [J]. J Anal Methods Chem 2014:409089
4. Li W, Hou M, Cao Y et al (2012) Determination of 20 free amino acids in asparagus tin by high-performance liquid chromatographic method after pre-column derivatization [J]. Food Anal Methods 5(1):62–68
5. Maria VFSD, Tatiana PC, Beatriz ASDM et al (2024) An improved method for determining free amino acids by RP-HPLC/DAD with o-phthalaldehyde derivatization: method evaluation in beers and wines [J]. Food Chem 435:137591–137591
6. Sabadel A, Browning T, Kruimer D et al (2017) Determination of picomolar dissolved free amino acids along a South Atlantic transect using reversed-phase high-performance liquid chromatography [J]. Mar Chem 196:173–180
7. Liyanaarachchi G, Mahanama K, Somasiri H et al (2018) Validation of a reversed-phase high-performance liquid chromatographic method for the determination of free amino acids in rice using l-theanine as the internal standard [J]. Food Chem 240:196–203
8. Min L, Daxiang L, Yuling T et al (2018) Determination of free amino acids in tea by a novel method of reversed-phase high performance liquid chromatography applying 6-aminoquinolyl- N-hydroxysuccinimidyl carbamate reagent [J]. J Food Sci Technol 55(10):4276–4286
9. Nie L, Ma S, Qian X et al (2021) Determination of free amino acids in banlangen granule and its fractions by solid phase extraction combined with ion-pair high performance liquid chromatography using a corona-charged aerosol detector (SPE-HPLC-CAD) [J]. Curr Pharm Anal 17(7):838–846
10. Albin MD, Wubben EJ, Gabert MV (2000) Effect of hydrolysis time on the determination of amino acids in samples of soybean products with ion-exchange chromatography or precolumn derivatization with phenyl isothiocyanate [J]. J Agricult Food Chem 48(5):1684–1691

Chapter 11

Determination of Total Amino Acids

Jinbo Guo and Zhangcheng Liang

Abstract

Protein is essential to human health and body growth. Currently, there is growing interest healthy and high nutritious foods, this drives food industry to develop high protein food product to meet this demand. Bread, a widely consumed staple food in the world, is a good protein vehicle for increasing consumers' daily protein intake. Notably, the nutritional quality of protein depends on their amino acid profile, especially in the compositions and content of essential amino acids. Therefore, determining the amino acid composition in bread is essential to evaluate the protein quality of the bread, which can further provide more nutritional information to consumer and guide bakery industry in bread reformulation and innovation. This chapter provide a detailed procedure for determination of total amino acid in bread using amino acid analyzer, following the Chinese national standards (GB 5009.124-2016).

Keywords Amino acid analysis, Bread, Food analysis, Nutritional quality, Protein

1 Introduction

Bakery products are widely consumed in the human diets worldwide, with bread representing one of the most important and popular staple foods [1, 2]. At the same time, bread constitutes a significant nutrients contributor in the human daily diet and is included in all dietary guidelines as part of a balanced diet [3, 4]. Due to the limitation of lysine content, most breads that made using refined wheat flour cannot provide high nutritional quality protein and are considered nutritionally inadequate [5, 6]. From a nutritional point of view, the nutritional quality of proteins depends largely on their amino acid composition, especially in the essential amino acids that the human body cannot synthesize and must obtain from daily diet [7, 8]. In order to improve the nutritional quality of the bread, many ingredients have been introduced into bread production to improve the amino acid balance [9–12]. In this regard, determining the amino acid content and profile in bread is essential to evaluate the nutritional quality of the final bread products [13]. This information not only provides more nutritional information to consumers, but also further

Guoqiang Zhang and Jinying Guo (eds.), *Frozen Dough and Bread Quality*, Methods and Protocols in Food Science,
https://doi.org/10.1007/978-1-0716-5206-0_11,

help bread industry in reformulating bread recipe to produce bread with balanced amino acid profile [14, 15].

Therefore, this chapter describes a detailed procedure for determining total amino acids in bread using amino acid analyzer (ion-exchange chromatography with postcolumn ninhydrin derivatization), following the procedures of the Chinese national standards (GB 5009.124-2016, determination of amino acid in foods) [16]. Proteins in bread are hydrolyzed by hydrochloric acid into free amino acids, which are separated by an ion-exchange column. The amino acids then react with ninhydrin solution to produce a color reaction, and their content is determined using a visible light spectrophotometric detector. This method applies to the determination of acid-hydrolyzed amino acids in bread, including aspartic acid, threonine, serine, glutamic acid, proline, glycine, alanine, valine, methionine, isoleucine, leucine, tyrosine, phenylalanine, histidine, lysine, and arginine—a total of 16 amino acids.

2 Materials

Unless otherwise specified, all reagents used in this method are of analytical grade, and the water complies with ultrapure water.

2.1 Reagents

1. Hydrochloric acid (HCl, concentration ≥36%).
2. Phenol (C_6H_5OH).
3. Nitrogen gas: Purity 99.9%.
4. Sodium citrate ($Na_3C_6H_5O_7{\cdot}2\,H_2O$).
5. Sodium hydroxide (NaOH).
6. Amino acid stands: Individual amino acid standards (solid, purity ≥98%) or mixed amino acid standard solution: A certified reference material with a national standard certificate.

2.2 Reagent Preparation

1. Hydrochloric acid solution (6 mol/L): Dilute 500 mL of hydrochloric acid with water to 1000 mL and mix well (*see* **Note 1**).
2. Freezing mixture: A 1:3 mass ratio mixture of commercially available salt and ice.
3. Sodium hydroxide solution (500 g/L): Dissolve 50 g of sodium hydroxide in 50 mL of water, cool to room temperature, dilute to 100 mL with water, and mix well.
4. Sodium citrate buffer solution [$c(Na^+)=0.2$ mol/L]: Dissolve 19.6 g of sodium citrate in 500 mL of water, add 16.5 mL of hydrochloric acid, dilute to 1000 mL with water, mix well, and adjust the pH to 2.2 using 6 mol/L hydrochloric acid solution or 500 g/L sodium hydroxide solution.

5. Elution buffer solutions with different pH and ionic strengths: Prepare according to the instrument manual or purchase commercially.
6. Ninhydrin solution: Prepare according to the instrument manual or purchase commercially (*see* **Note 2**).

2.3 Apparatus Required

1. Laboratory tissue grinder or mill.
2. Homogenizer.
3. Analytical balance: With sensitivities of 0.0001 and 0.00001 g.
4. Hydrolysis tube: Pressure-resistant screw-cap glass tube or ampoule, volume 20 mL–30 mL.
5. Vacuum pump: Exhaust capacity ≥40 L/min.
6. Alcohol blast lamp.
7. Electric thermostatic drying oven or hydrolysis furnace.
8. Tube concentrator or parallel evaporator (with matching 15–25 mL tubes).
9. Amino acid analyzer: Ion-exchange chromatograph with post-column ninhydrin derivatization (Hitachi, L-8900) (*see* **Note 3**).

3 Methods

3.1 Standard Solution Preparation

1. Mixed amino acid standard stock solution (1 μmol/mL): Accurately weigh each individual amino acid standard (to 0.00001 g) into the same 50 mL beaker, dissolve with 8.3 mL of 6 mol/L hydrochloric acid solution, transfer quantitatively to a 250 mL volumetric flask, dilute to the mark with water, and mix well (refer to Table 1 for the reference mass of each amino acid standard).
2. Mixed amino acid standard working solution (100 nmol/mL): Accurately pipette 1.0 mL of the mixed amino acid standard stock solution into a 10 mL volumetric flask, dilute to the mark with pH 2.2 sodium citrate buffer solution, mix well, and use as the standard injection solution.

3.2 Analytical Procedure

3.2.1 Sample Preparation

Bread samples are ground using a tissue grinder or mill, while liquid samples are homogenized using a homogenizer, sealed, and stored frozen. Thaw before analysis.

3.2.2 Sample Hydrolysis

1. Add 10 mL of 6 mol/L hydrochloric acid solution to the hydrolysis tube and 3–4 drops of phenol to the hydrolysis tube (*see* **Note 4**).

Table 1
Reference values for the weighing mass of amino acid standards and their molecular weights when preparing mixed amino acid standard stock solutions

Amino acid standard name	Weighing reference (mg)	Molar mass (g/moL)	Amino acid standard name	Weighing reference (mg)	Molar mass (g/moL)
L-aspartic acid	33	133.1	L-methionine	37	149.2
L-threonine	30	119.1	L-isoleucine	33	131.2
L-serine	26	105.1	L-leucine	33	131.2
L-glutamic acid	37	147.1	L-tyrosine	45	181.2
L-proline	29	115.1	L-phenylalanine	41	165.2
Glycine	19	75.07	L-histidine hydrochloride	52	209.7
L-alanine	22	89.06	L-lysine hydrochloride	46	182.7
L-valine	29	117.2	L-arginine hydrochloride	53	210.7

2. Place the hydrolysis tube in a freezing mixture for 3–5 min, connect to a vacuum pump, evacuate (close to 0 Pa), and then fill with nitrogen. Repeat the evacuation-nitrogen filling process three times, then seal or tighten the screw cap under nitrogen (*see* **Note 5**).
3. Place the sealed hydrolysis tube in an electric thermostatic drying oven or hydrolysis furnace at 110°C ± 1°C for 22 h. Remove and cool to room temperature.
4. Open the hydrolysis tube, filter the hydrolysate into a 50 mL volumetric flask, rinse the tube repeatedly with small amounts of water, transfer the rinsate to the same flask, and dilute to the mark with water. Mix well.
5. Pipette 1.0 mL of the filtrate into a 15 or 25 mL tube, and dry under reduced pressure at 40–50°C using a tube concentrator or parallel evaporator. Dissolve the residue in 1–2 mL of water, dry again, and evaporate to dryness.
6. Add 1.0–2.0 mL of pH 2.2 sodium citrate buffer solution to the dried tube, mix well, filter through a 0.22 μm of membrane, transfer to an instrument sample vial, and use as the sample test solution for instrumental analysis in amino acid analyzer.

3.3 Amino Acid Analyzer Procedure and Conditions

Inject the mixed amino acid standard working solution into the amino acid analyzer. Adjust the instrument operation program, parameters, and elution buffer solution ratios according to JJG 1064—2011 verification regulation for amino acid analyzers and the instrument manual to confirm the operating conditions. The chromatographic reference conditions are listed below:

(a) Column: Sulfonic acid-type cation-exchange resin.

(b) Detection wavelengths: 570 and 440 nm.

3.4 Sample Determination

Inject equal volumes of the mixed amino acid standard working solution and sample test solution into the amino acid analyzer. Calculate the concentration of amino acids in the sample test solution using the external standard method based on peak areas.

3.5 Expression of Results

3.5.1 Calculation of Amino Acid Concentrations in Mixed Amino Acid Standard Stock Solution

The concentration of amino acid in the mixed amino acid standard stock solution is calculated using Eq. (1):

$$cj = \frac{mj}{Mj \times 250} \times 1000 \quad (1)$$

c_j—Concentration of amino acid j in the mixed amino acid standard stock solution, in micromoles per milliliter (μmol/mL).
m_j—Mass of amino acid standard *j* weighed, in milligrams (mg).
M_j—Molecular weight of amino acid standard *j*.
250—Volumetric flask volume, in milliliters (mL).
1000—Conversion factor.
Results are retained to four significant figures.

3.5.2 Calculation of Amino Acid Content in Samples

The content of amino acid in the sample test solution is calculated using Eq. (2):

$$ci = \frac{cs}{As} \times Ai \quad (2)$$

c_i—Content of amino acid i in the sample test solution, in nanomoles per milliliter (nmol/mL).
Ai—Peak area of amino acid *i* in the sample test solution.
As—Peak area of amino acid *s* in the standard working solution.
c_s—Content of amino acid *s* in the standard working solution, in nanomoles per milliliter (nmol/mL).

The content of each amino acid in the sample is calculated using Eq. (3):

$$Xi = \frac{ci \times F \times V \times M}{\mathrm{m} \times 10^9} \quad (3)$$

Table 2
Names and molar masses of the 16 amino acids

Amino acid name	Molar mass (g/moL)	Amino acid name	Molar mass (g/moL)
Aspartic acid	133.1	Methionine	149.2
Threonine	119.1	Isoleucine	131.2
Serine	105.1	Leucine	131.2
Glutamic acid	147.1	Tyrosine	181.2
Proline	115.1	Phenylalanine	165.2
Glycine	75.1	Histidine	155.2
Alanine	89.1	Lysine	146.2
Valine	117.2	Arginine	174.2

For amino acid contents below 1.00/100 g, retain two significant figures; for contents above 1.00/100 g, retain three significant figures.

Xi—Content of amino acid *i* in the sample, in grams per hundred grams (g/100 g).
ci—Content of amino acid *i* in the sample test solution, in nanomoles per milliliter (nmol/mL).
F—Dilution factor.
V—Volume of the hydrolysate transferred and diluted, in milliliters (mL).
M—Molar mass of amino acid *i*, in grams per mole (g/mol).

Refer to Table 2 for the names and molar masses of the amino acids.
m — Sample mass, in grams (g).
10^9—Conversion factor from nanograms (ng) to grams (g).
100—Conversion factor.

4 Notes

1. Hydrochloric acid is a corrosive substance; therefore, personal protective equipment is required during the preparation.
2. This formula needs to be prepared on the spot and is prone to decomposition and failure if stored for a long time.
3. Nitrogen pressure: The instrument requires a pressure of 0.05–0.1 Mpa, and excessive pressure can damage the instrument. Besides, the injection concentration ranges from 1 to

100 nmol/mL, excessive concentrations can affect the analysis results and easily block the reaction column.

4. Perform operations in a fume hood under closed conditions, ensuring the exhaust system is functioning effectively.
5. Need to ensure that there is no oxygen in the hydrolysis tube.

References

1. Zhang G, Li Z (2024) Impact of melon seed oil cake with different particle sizes on bread quality. FPPN 6:44. https://doi.org/10.1186/s43014-024-00225-9
2. Nudel A, Cohen R, Abbo S, Kerem Z (2023) Developing a nutrient-rich and functional wheat bread by incorporating *Moringa oleifera* leaf powder and gluten. LWT 187:115343. https://doi.org/10.1016/j.lwt.2023.115343
3. Lockyer S, Spiro A (2020) The role of bread in the UK diet: an update. Nutr Bull 45:133–164. https://doi.org/10.1111/nbu.12435
4. Dong Y, Karboune S (2021) A review of bread qualities and current strategies for bread bioprotection: flavor, sensory, rheological, and textural attributes. Compr Rev Food Sci Food Saf 20:1937–1981. https://doi.org/10.1111/1541-4337.12717
5. Montevecchi G, Santunione G, Licciardello F, Köker Ö, Masino F, Antonelli A (2022) Enrichment of wheat flour with Spirulina. Evaluation of thermal damage to essential amino acids during bread preparation. Food Res Int 157:111357. https://doi.org/10.1016/j.foodres.2022.111357
6. Tsegay N, Admassu H, Zegale B, Gosu A (2024) Nutritional and functional potentials of wheat, cowpea, and yam composite flours on bread formulations: effect of blending ratio and baking parameters. J Agric Food Res 18:101294. https://doi.org/10.1016/j.jafr.2024.101294
7. Igual M, García-Segovia P, Martínez-Monzó J (2021) Amino acids release from enriched bread with edible insect or pea protein during in vitro gastrointestinal digestion. Int J Gastron Food Sci 24:100351. https://doi.org/10.1016/j.ijgfs.2021.100351
8. Zhang G, Wang Y, Liang Z, Guo J, Wu B (2025) Nutritional value and advances of mealworm (*Tenebrio molitor*) as a sustainable functional ingredient in bakery products: a key step toward sustainable diets. FF 12:100716. https://doi.org/10.1016/j.fufo.2025.100716
9. Ferreyra LS, Verdini RA, Soazo M, Piccirilli GN (2021) Impact of whey protein addition on wheat bread fermented with a spontaneous sourdough. Int J Food Sci Technol 56:4738–4745. https://doi.org/10.1111/ijfs.15052
10. Zhang G, Chatzifragkou A, Charalampopoulos D, Rodriguez-Garcia J (2023) Effect of defatted melon seed residue on dough development and bread quality. LWT 183:114892. https://doi.org/10.1016/j.lwt.2023.114892
11. Rusu IE, Marc Vlaic RA, Mureşan CC, Mureşan AE, Mureşan V, Pop CR, Chiş MS, Man SM, Filip MR, Onica B-M, Alexa E, Vişan V-G, Muste S (2021) Hemp (*Cannabis sativa* L.) flour-based wheat bread as fortified bakery product. Plants 10:1558. https://doi.org/10.3390/plants10081558
12. Khan MKI, Asif M, Razzaq ZU, Nazir A, Maan AA (2022) Sustainable food industrial waste management through single cell protein production and characterization of protein enriched bread. Food Biosci 46:101406. https://doi.org/10.1016/j.fbio.2021.101406
13. Biduski B, Maçãs M, Vahedikia N, O'Connor PM, Hussey K, Simpson JC, Mysior MM, Gallagher E (2024) Dough rheology and internal structure of bread produced with wheat flour partially substituted by buckwheat flour: a step towards enhancing nutritional value. Food Struct 39:100364. https://doi.org/10.1016/j.foostr.2023.100364
14. Zhang G, Guo J, Guo J (2025) A sustainable approach in pumpkin seed oil processing line: recent advances in pumpkin seed oil and oil processing by-products. Food Chem X 26:102259. https://doi.org/10.1016/j.fochx.2025.102259
15. Prieto-Vázquez del Mercado P, Mojica L, Morales-Hernández N (2022) Protein ingredients in bread: technological, textural and health implications. Foods 11:2399. https://doi.org/10.3390/foods11162399
16. China Food and Drug Administration, National Food Safety Standard-Determination of Amino Acid in Foods (2016). https://www.codeofchina.com/standard/GB5009.124-2016.html

Check for updates

Chapter 12

Element Analysis

Guoqiang Zhang

Abstract

Bread, one of popular staple foods in the world, is considered an important source of minerals in daily diet. However, from bread production to sale, heavy metals could enter bread products and adversely affect food safety and consumer health. With the increasing health awareness of consumers, the quantification of these elements in bread products is essential to ensure food quality and safety. Inductively coupled plasma optical emission spectroscopy (ICP-OES) is one of widely techniques for elemental determination in food samples, which has high sensitivity and accuracy. Therefore, this chapter describes a detailed procedure that is employed for element analysis of bread products using ICP-OES. The detailed protocol is divided into two parts, including sample digestion and determination of element content by ICP-OES.

Keywords Bread, Bread nutritional value, Mineral content determination, Element analysis, Food analysis, Inductively coupled plasma optical emission spectroscopy (ICP-OES)

1 Introduction

Minerals are one of the important nutrients for essential functions of the body and the maintenance of human health such as the synthesis of enzymes, immune function, fluid balance, and bone health [1]. Notably, minerals deficiency in a diet with adequate energy intake is considered hidden hunger, which may result in poor health [2]. According to World Health Organization (WHO) estimates, hidden hunger affects over two billion people worldwide, mainly due to deficiencies in microminerals such as iron and zinc [3].

Bread, as one of the popular staple foods worldwide, provides significant contributors to human nutrients daily intake, such as fiber, protein, and minerals (e.g., potassium, calcium, zinc, magnesium, and iron) [4, 5]. Based on this, bread is a feasible delivery vehicle of nutrients. Many studies have reported that incorporation of other nutrient-rich ingredients into bread formulations is a promising method to enhance bread nutritional quality and is considered as one of the most effective interventions to address microminerals deficiencies of the population [6–8]. The mineral content in bread

Guoqiang Zhang and Jinying Guo (eds.), *Frozen Dough and Bread Quality*, Methods and Protocols in Food Science, https://doi.org/10.1007/978-1-0716-5206-0_12,

depends on multiple factors such as ingredients used, processing methods, and the geographical position of crops [9]. On the other hand, from food safety perspective, ingredients contaminated with heavy metals (e.g., lead, nickel, and cadmium) could be used in the bread production, resulting in food fraud and posing a serious risk to customer's health [10, 11]. Additionally, there is the possibility of exposure to heavy metal during bread production to sale, leading to contamination [11]. Therefore, determining the content of dietary elements and harmful elements in bread are of great significant for evaluating food quality and ensuring food safety.

Inductively coupled plasma optical emission spectroscopy (ICP-OES) is one of widely used techniques for element analysis in various food products, since it has high sensitivity and accuracy, easy operation, and ability to analyze multielement simultaneously (2–74 elements) [12–14]. This technique uses inductively coupled plasma as an excitation source, and then detects and measures the characteristic emission spectra of elements in samples [13]. Notably, ICP-OES is a technique developed for liquids; therefore, most of solid samples should be digested by acids (nitric acid, sulfuric acid, and hydrochloric acid) at high temperature [15]. Nitric acid (HNO_3) is commonly used because of its oxidizing properties and ability to minimize the formation of insoluble salts; in addition, sulfuric acid and hydrochloric acid can be added along with nitric acid to enhance sample digestion if nitric acid along does not achieve complete digestion [12, 15]. Besides, hydrogen peroxide (H_2O_2) can be combined with acids to increase the oxidant power and improve digestion efficiency [16, 17].

This chapter describes a detailed procedure for the elemental determination of bread samples using ICP-OES technique, including sample digestion procedures using the HNO_3 with H_2O_2 and element content determination using ICP-OES. Briefly, bread sample is digested using HNO_3 and H_2O_2 at high temperature. After digestion, the final digests are filtered, and then subjected to elemental determination using ICP-OES.

2 Materials

2.1 Reagents

All chemical reagents are of analytical grade unless otherwise stated.

1. Nitric acid (65%, w/w).
2. Deionized water or HPLC-grade water (optional).
3. Hydrogen peroxide (30%, w/w).
4. Element standard solutions (1000 mg/L).

2.2 Apparatus Required

Soak all glassware in 10% (v/v) nitric acid solution for 24 h and rinse repeatedly with deionized water before use.

1. Beakers.
2. Analytical balance.
3. Volumetric flasks (50 mL).
4. Hot plate–set at 100°C.
5. Pipettes.
6. Filter papers (Whatman No. 1).
7. Watch glass.
8. Funnel.
9. ICP-Optical Emission Spectrometer (Optima 8300, PerkinElmer, USA).

3 Methods

Follow all waste disposal regulations when disposing of waste materials. Follow the safety regulations, wear recommended personal protective equipment (PPE).

3.1 Sample Digestion

The procedure is illustrated in Fig. 1.

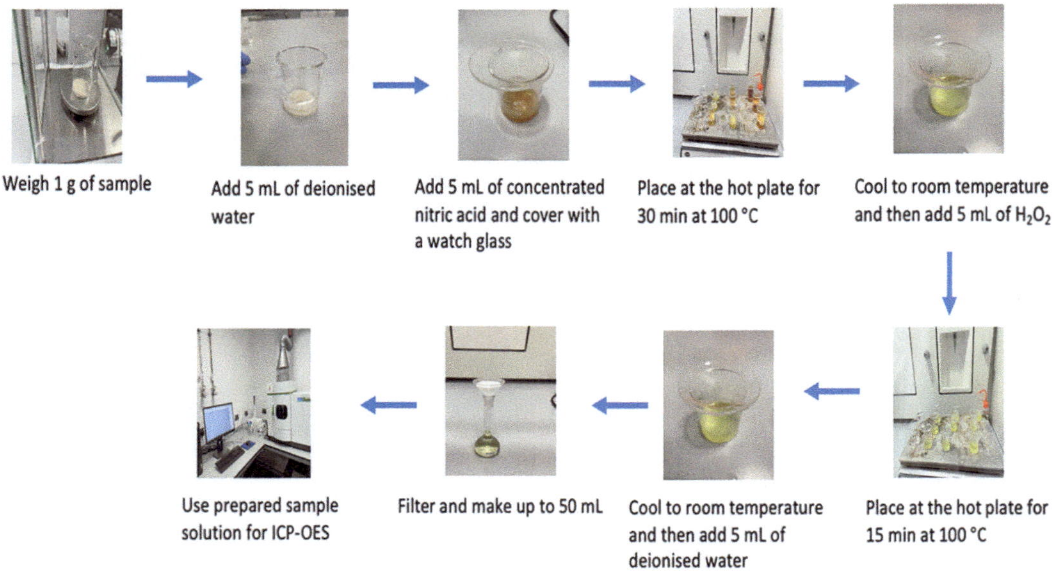

Fig. 1 Simplified schematic representation of bread sample digestion procedure for ICP-OES

1. Weigh 1 g of sample in the 50 mL of beaker, record the exact weight of the sample to the nearest 0.1 mg (*see* **Notes 1 and 2**).
2. Add 5 mL of deionized water in the beaker.
3. Add 5 mL of concentrated HNO_3 (65%, w/w) in the beaker and cover the beaker with a watch glass (*see* **Note 3**).
4. Transfer the beaker to a hot plate at 100°C for 30 min (*see* **Note 4**).
5. After 30 min, remove the beaker from the hot plate and cool to room temperature for 10 min (*see* **Note 5**).
6. Add 5 mL of H_2O_2 (30%, w/w) in the beaker and place the beaker to the hot plate for 15 min at 100°C.
7. After 15 min, remove the beaker from the hot plate and allow to cool to room temperature for 10 min (*see* **Note 5**).
8. Add 5 mL of deinoized water in the beaker and quantitatively transfer the solution through a Whatman No. 1 filter paper to make up to 50 mL, and then mix thoroughly (*see* **Notes 6 and 7**).
9. Use the prepared solution to determine mineral content using inductively coupled plasma optical emission spectroscopy (ICP-OES).

3.2 Mineral Content Determination Using ICP-OES

1. Prepare a series concentration ranges of external multielement standard solutions using 2% nitric acid solution, with concentration expressed in mg/L (*see* **Notes 8 and 9**).
2. The ICP-OES operating conditions are shown in Table 1 (*see* **Note 10**).
3. Set the readout to nil using 2% nitric acid solution (as calibration blank), and then analyze external multielement standard solutions (*see* Fig. 2 and **Note 11**).
4. Analyze the blank sample at first and then for the prepared samples (*see* **Notes 11 and 12**).
5. Syngistix ICP software (version 2.0) is used to control system and analyze the data.

3.3 Calculation

Use Equation to determine each element content in the sample:

$$\mathrm{W(mg/100g\ of\ sample)} = \frac{Cs\ x\ V}{1000x\ Ws} \mathrm{x\ F\ x100}$$

where,

W = the calculated element in the tested sample, expressed as mg/100 g.

C_s = the concentration of sample solution (mg/L).

Table 1
Operating parameters used for mineral content determination by ICP-OES

Parameter	ICP-OES
Reflect power (W)	1500
Plasma gas flow rate (L/min)	8.0
Auxiliary gas flow rate (L/min)	0.2
Nebulizer gas flow rate (L/min)	0.7
Pump flow rate (mL/min)	1.0
Sample flow rate (mL/min)	1.0
Replicates	3
Element wavelength (nm)	
Potassium (K)	766.490
Magnesium (Mg)	285.213
Calcium (Ca)	317.933
Aluminum (Al)	396.153
Iron (Fe)	238.204
Copper (Cu)	327.393
Zinc (Zn)	206.200
Sliver (Ag)	328.068
Manganese (Mn)	285.213
Sodium (Na)	589.592
Nickel (Ni)	231.604
Lead (Pb)	220.353

V = volume of sample after (50 mL).
W_s = sample weight (g).
F = dilution factor; further dilution of sample solution (either undiluted).

4 Notes

1. Solid sample is typically dried to eliminate the influence of varying moisture content. Besides, to ensure that the target analytes are evenly distributed in sample, it is necessary to grind, sieve, and mix sample after drying process.

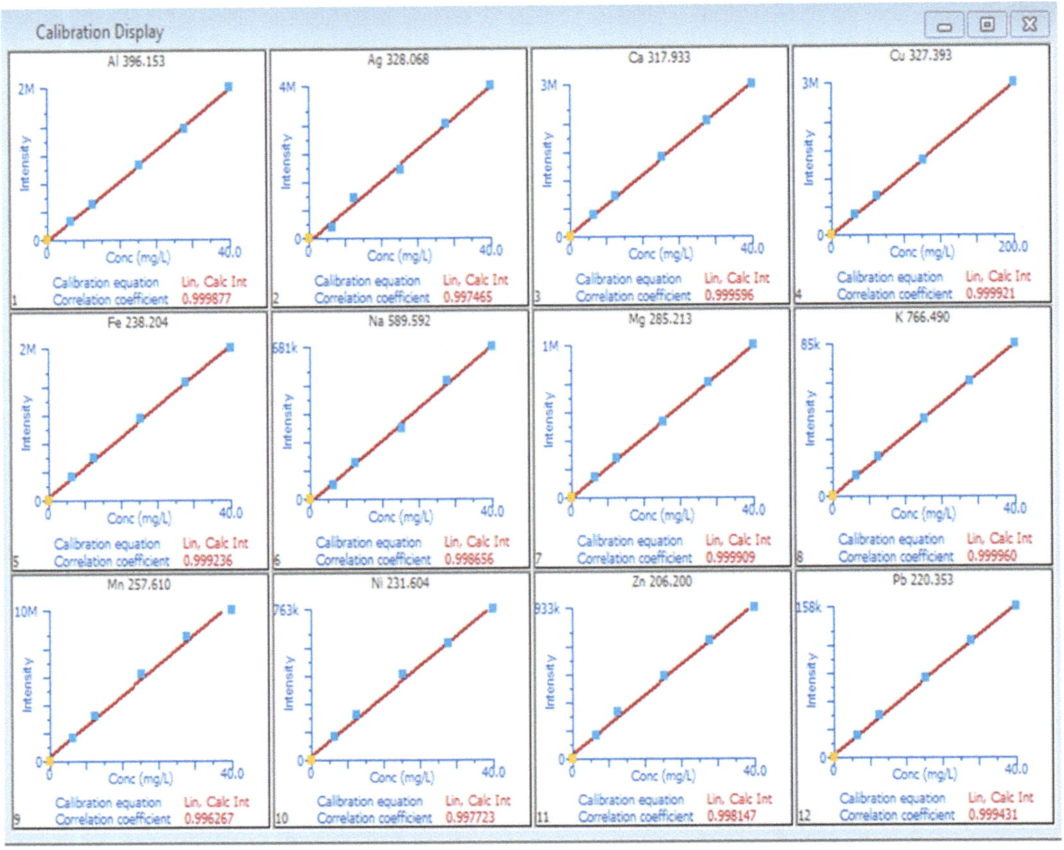

Fig. 2 An example of calibration curves for 12 elements and operating wavelengths (*see* **Note 9**)

2. Prepare one empty beaker without any sample as a blank, following the same procedure and adding the same amount of each reagent for sample digestion steps.
3. The corrosive reagent, i.e., nitric acid (HNO_3), is used in this experiment. To ensure the safety of operators, it is necessary to wear personal protective equipment and operate it in the fume cupboard.
4. Regularly check the heating status in each 5 min and gently shake for 5 s to prevent boiling dry and ensure the contents are evenly heated.
5. The beaker is extremely hot and contains corrosive reagent, it is necessary to handle it carefully with gloves.
6. Prewetting the filter paper with a small amount of deinoized water after placing it in the funnel will help it to spread properly in the funnel and create a good seal.
7. Rinse the beaker 3–5 times using small amount of deionized water to ensure all substance is transferred to the volumetric flask.

8. A series concentration ranges of multielement standard solutions can be prepared using single-element standard solutions or multielements standard solutions.
9. The concentration ranges of standard solutions depend on the tested sample. Generally, bread products contain low level of microminerals such as copper (Cu), zinc (Zn), and iron (Fe), but this also depends on the bread formulation and ingredient used. Therefore, multiple tests are required to determine the appropriate concentration ranges of standard solutions for the tested sample.
10. The machine needs to warm up at least 60 min before using.
11. Rinse with 2% nitric acid solution between sample measurements.
12. It is important to note that the solutions contain larger particulates that may cause blockage; therefore, the solutions should be clear. Centrifuge the solution before analysis, if necessary.

References

1. Barone M, D'Amico F, Brigidi P, Turroni S (2022) Gut microbiome–micronutrient interaction: the key to controlling the bioavailability of minerals and vitamins? Biofactors 48:307–314
2. Lowe NM (2021) The global challenge of hidden hunger: perspectives from the field. Proc Nutr Soc 80:283–289
3. World Health Organization. Preventing and controlling micronutrient deficiencies in populations affected by an emergency [Internet]. 2006 [cited 2025 Mar 2]. https://cdn.who.int/media/docs/default-source/nutritionlibrary/preventing-and-controlling-micronutrient-deficiencies-in-populations-affected-by-an-emergency.pdf?sfvrsn=e17f6dff_4&download=true
4. Ribet L, Kassis A, Jacquier E, Monnet C, Durand-Dubief M, Bosco N (2025) The nutritional contribution and relationship with health of bread consumption: a narrative review. Crit Rev Food Sci Nutr 65(28):5698–5725. https://doi.org/10.1080/10408398.2024.2428593
5. Zhang G, Chatzifragkou A, Charalampopoulos D, Rodriguez-Garcia J (2023) Effect of defatted melon seed residue on dough development and bread quality. LWT 183:114892
6. Issaoui M, Oueslati S, Flamini G, Delgado AM, Romano A (2024) Fortification of wheat white bread: assessing the suitability of *Beta vulgaris* trough technological, nutritional, and sensory evaluation. Int J Food Sci Technol 59:8755–8768
7. Cormick G, Betran A, Romero I, Cormick M, Belizán J, Bardach A et al (2021) Effect of calcium fortified foods on health outcomes: a systematic review and meta-analysis. Nutrients 13:316
8. Zhang G, Li Z (2024) Impact of melon seed oil cake with different particle sizes on bread quality. FPPN 6:44
9. Moisa C, Brata AM, Muresan IC, Dragan F, Ratiu I, Cadar O et al (2024) Comparative analysis of vitamin, mineral content, and antioxidant capacity in cereals and legumes and influence of thermal process. Plants 13:1037
10. Khursheed MH ul R, Shahbaz M, Ramzan T, Haider A, Maqsood MF, Khan A et al (2024) Enhancing wheat tolerance to cadmium stress through moringa leaf extract foliar application. Scientifica (Cairo) 2024(1):2919557. https://doi.org/10.1155/2024/2919557
11. Zioła-Frankowska A, Karaś K, Mikołajczak K, Kurzyca I, Kowalski A, Frankowski M (2021) Identification of metal(loid)s compounds in fresh and pre-baked bread with evaluation of risk health assessment. J Cereal Sci 97:103164
12. Ncube N, Tancu Y, Mketo N (2024) A greener, rapid and accurate microwave-assisted hydrogen peroxide digestion method for ICP-OES determination of heavy metals in pet food samples. J Food Comp Anal 131:106201
13. Rawat H, Bhat SA, Dhanjal DS, Singh R, Gandhi Y, Mishra SK et al (2024) Emerging

techniques for the trace elemental analysis of plants and food-based extracts: a comprehensive review. Talanta Open 10:100341
14. Fontoura BM, Jofré FC, Williams T, Savio M, Donati GL, Nóbrega JA (2022) Is MIP-OES a suitable alternative to ICP-OES for trace element analysis? J Anal At Spectrom 37:966–984
15. Douvris C, Vaughan T, Bussan D, Bartzas G, Thomas R (2023) How ICP-OES changed the face of trace element analysis: review of the global application landscape. Sci Total Environ 905:167242
16. Zuma MC, Nomngongo PN, Mketo N (2021) Simultaneous determination of REEs in coal samples using the combination of microwave-assisted ashing and ultrasound-assisted extraction methods followed by ICP-OES analysis. Minerals 11:1103
17. Santos HM, Coutinho JP, Amorim FAC, Lôbo IP, Moreira LS, Nascimento MM et al (2019) Microwave-assisted digestion using diluted HNO3 and H2O2 for macro and microelements determination in guarana samples by ICP OES. Food Chem 273:159–165

Check for updates

Chapter 13

Evaluating Phenolic Compounds and Antioxidant Properties

Jia Wei

Abstract

Phenolic compounds (PCs) are a structurally diverse group of bioactives in cereal-based systems, contributing significantly to antioxidant potential, sensory quality, and nutritional value. Their profiles are influenced by grain type, flour pigmentation, fermentation, and processing conditions. During dough fermentation and baking, PCs undergo enzymatic and microbial modifications that enhance their solubility and bioavailability. In frozen dough systems, physical stress during freezing and thawing can also facilitate the release of bound PCs, while prolonged storage may lead to degradation. Accurate characterization of PCs requires effective extraction methods tailored to food matrices, along with antioxidant assays. This chapter describes common techniques for extracting free and bound phenolics, as well as widely used in vitro assays including Folin–Ciocalteu, DPPH$^{\cdot}$, ABTS$^{\cdot+}$, FRAP, and CUPRAC. These colorimetric methods, often performed in microplate or cuvette formats, are essential for assessing the effects of formulation and processing on PC composition and antioxidant activity in frozen dough and bread.

Keywords Phenolic compounds, Frozen dough, Bread, Antioxidant activity, Extraction

1 Introduction

Phenolic compounds (PCs) represent a structurally diverse group of plant-derived secondary metabolites, defined by the presence of one or more hydroxyl groups attached to aromatic rings. This group encompasses a broad array of subclasses, including phenolic acids, flavonoids, tannins, lignans, curcuminoids, coumarins, carotenoids, stilbenes, and quinones. Widely distributed in plant-based foods such as cereals, legumes, fruits, and vegetables, PCs occur in two primary forms: soluble phenolics, which are generally compartmentalized within vacuoles, and bound phenolics, which are covalently associated with cell wall macromolecules—primarily cellulose, hemicellulose, and proteins—through ester or glycosidic linkages. In cereal-based systems, PCs were historically considered antinutritional due to their propensity to form complexes with dietary proteins and inhibit digestive enzymes, thereby diminishing nutrient bioavailability and imparting undesirable sensory attributes, such as bitterness and astringency [1].

Guoqiang Zhang and Jinying Guo (eds.), *Frozen Dough and Bread Quality*, Methods and Protocols in Food Science,
https://doi.org/10.1007/978-1-0716-5206-0_13,

However, this perception has shifted significantly in recent years. An expanding body of evidence highlights the multifunctional role of PCs beyond their antinutritional effects. They are now recognized as precursors of key volatile aroma compounds, thereby enhancing the flavor complexity of baked products. Moreover, their potent antioxidant, anti-inflammatory, anticancer, and antimicrobial activities have elevated their status as essential bioactive components, contributing to the health-promoting potential of cereal-based foods.

Whole cereal grains exhibit substantial variability in their phenolic profiles, both in composition and concentration. Wheat grains are primarily composed of phenolic acids, most of which are present in the bound form, thereby limiting their bioavailability during digestion. Oats are notable for their high content of avenanthramides, a unique class of phenolic alkaloids with well-documented antioxidant and anti-inflammatory properties. Barley contains a broader spectrum of phenolic constituents, including phenolic acids, flavonoids, and lignans, while rye is particularly rich in ferulic acid and its related compounds [2]. Pigmented cereal grains, such as colored corn varieties, are especially abundant in anthocyanins, which contribute not only to their distinctive coloration but also to their potential health-promoting effects through strong radical-scavenging capacity. In addition, some cereal grains contain rare and taxonomically specific phenolics; for example, sorghum accumulates 3-deoxyanthocyanins, a subclass of anthocyanins not typically found in other grains. With respect to total phenolic content, corn generally ranks the highest among major cereals, with concentrations reaching approximately 15.55 μg/g. This is followed by wheat, oats, and rice, which exhibit moderate but still nutritionally relevant levels, averaging around 5.56 μg/g [3].

In frozen dough and bread systems, PCs play a particularly crucial role due to the structural and chemical transformations they undergo during freezing, thawing, and subsequent baking. The formation of ice crystals and the associated mechanical stresses during freezing can disrupt cell walls and starch-protein matrices, thereby facilitating the release of bound phenolics into their free forms. This release can enhance the antioxidant capacity of the system by increasing the availability of redox-active compounds. However, these benefits are counterbalanced by potential losses: during frozen storage, prolonged exposure to oxygen, light, and temperature fluctuations may induce oxidative degradation, phenolic–aldehyde polymerization, and structural modifications that reduce the biological activity of PCs. Such changes are particularly pronounced in thermolabile and oxygen-sensitive subclasses, including flavonols and anthocyanins.

Dough fermentation is known to facilitate the release of bound phenolic compounds from the cereal matrix during fermentation. As a common strategy in frozen dough systems, dough fermentation not only improves the nutritional and sensory properties of the final product but also significantly influences the profile and bioavailability of PCs [4]. In quinoa, fermentation with a mixed culture of *L. actobacillus*, *Lc. casei*, and *Lc. paracasei* results in significantly higher concentrations of procyanidin B2, quercetin, and kaempferol compared to nonfermented controls. In red sorghum sourdough, strains such as *Lc. plantarum*, *Lm. Reuteri*, or *Lm. fermentum* exhibit glycosidase activity, releasing flavanone aglycones from glycosylated precursors. Hydroxycinnamic acid esterase in *L. johnsonii*, *L. acidophilus*, and *L. reuteri*, contributes to an increase free hydroxycinnamic acid content in whole-grain barley and oat groats. *L. rhamnosus* may decrease ferulic acid levels through further bioconversion into downstream metabolites. These microbial transformations collectively influence the antioxidant potential and nutritional value of frozen dough and bread [5–7].

The interactions between proteins and PCs in cereal-based systems are influenced by factors such as pH, temperature, ionic strength, protein structure, and the physicochemical characteristics of phenolics, including molecular weight, hydroxylation, methylation, glycosylation, and structural flexibility [8]. These interactions could be covalent or noncovalent, with the latter—primarily hydrogen bonding and hydrophobic interactions—playing a dominant role in reversible binding. Covalent interactions are typically formed through oxidation reactions, often altering protein conformation. Wheat gluten has been shown to interact with phenolic acids such as caffeic, ferulic, syringic, and gallic acids, as well as with tannins, via both covalent and noncovalent mechanisms [9]. Phenolic addition may influence protein folding, especially tertiary structure, without necessarily affecting secondary structure. During dough mixing, gluten networks are formed through disulfide bonds, hydrogen bonds, and hydrophobic interactions, which may be further modified by phenolics [10]. Starch, alongside gluten, is a major component of cereal-based systems, and its behavior is also affected by phenolics. During heating, gluten and starch compete for water, and gluten may limit starch gelatinization by forming a barrier around granules. Hydrophilic phenolics can interfere with this process by binding water or forming noncovalent bonds with starch, which may shift system pH and alter gelatinization properties [11]. Such interactions can lead to increased gelatinization temperatures and modified pasting behavior.

In recent years, increasing attention has been paid to the role of phenolic compounds in enhancing the nutritional and functional quality of frozen dough and bread. Among the phenolic acids present in wheat, ferulic acid is the most abundant. PCs are predominantly concentrated in the outer layers of the grain; therefore, their levels are particularly high in wholegrain and bran-rich products. Colored cereals such as black wheat, red wheat, black rice, and red sorghum have been shown to contain significantly higher levels of total phenolic content (TPC) and total flavonoid content (TFC), along with superior antioxidant capacity. In comparison to conventional yellow wheat, black wheat exhibits a six-fold higher TPC, along with greater levels of protein, total flavonoids, and TFC, and antioxidant activity [12]. Incorporating these pigmented grains into frozen dough and bread formulations can not only improve nutritional value and antioxidant functionality but also impart distinctive color, flavor, and market appeal. Additionally, the inclusion of natural functional ingredients, such as green tea powder, turmeric powder, grape seed extracts, etc., further enriches the phenolic composition, enhances textural properties, extends shelf life, and contributes to the overall quality and health-promoting potential of dough and bread [8, 13].

This chapter presents an overview of the most commonly applied analytical methods for evaluating phenolic compounds in dough and bread, as well as their associated antioxidant activity. It begins with a description of the procedures used to extract both free and bound phenolics from bread matrices. This is followed by detailed descriptions of several widely used antioxidant assays. These include the Folin–Ciocalteu method, the $DPPH^{\cdot}$ and $ABTS^{\cdot+}$ radical scavenging assays, the ferric reducing antioxidant power (FRAP) assay, and the cupric ion reducing antioxidant capacity (CUPRAC) assay. Each of these methods can be performed using either microplate or cuvette-based formats.

The investigation of PCs in frozen dough and bread systems involves key steps such as sample preparation, extraction, and purification. Pretreatment methods like freeze-drying are often used to preserve sample integrity and aid analysis, though care is needed to avoid changes in phenolic profiles. Extraction is a critical step affecting both recovery and representativeness, especially in complex food matrices. Free phenolics are typically extracted with aqueous ethanol or methanol, while bound phenolics, linked to cell wall components or gluten structures, require alkaline or acid hydrolysis for effective release. The ratio of free to bound phenolics varies with formulation, fermentation, and baking conditions, making extraction method and solvent choice essential. Factors such as pH, temperature, and duration also influence compound stability, as high temperatures could degrade sensitive phenolics. While traditional methods like Soxhlet extraction are still used, newer techniques such as ultrasound-assisted and pressurized liquid extraction offer improved efficiency and better preservation of phenolic functionality [3, 14]. The choice of extraction method must be tailored to the dough or bread matrix to accurately assess PCs composition and bioactivity within these processed cereal systems.

The chemometrics-powered spectroscopic techniques, such as infrared (IR), Fourier-transform near-infrared (FT-NIR), mid-infrared (MIR), ultraviolet–visible (UV–Vis), fluorescence, and Raman spectroscopy, have become widely used for the rapid and nondestructive quantification of phenolic compounds in both plant-based and animal-based food matrices. These methods offer high-throughput capabilities and require minimal sample preparation, making them attractive for industrial and research applications. However, despite the advancement of these modern tools, the Folin–Ciocalteu assay remains the most widely accepted and traditionally used method for the determination of TPC, particularly in baked products such as bread. This assay is based on single electron transfer (SET) reactions between the Folin–Ciocalteu reagent and phenolic compounds. At an alkaline pH (around 10), the deprotonation of phenolic hydroxyl groups leads to the formation of phenolate ions, which reduce the yellow Folin–Ciocalteu reagent to a blue-colored complex, measurable by spectrophotometry. Nevertheless, the assay is not entirely specific to phenolics, as other reducing agents—such as reducing sugars, ascorbic acid, and sulfites—can also participate in the redox reaction, potentially leading to overestimation of phenolic content [15]. Despite these limitations, the Folin–Ciocalteu method continues to serve as a benchmark in phenolic analysis due to its simplicity, reproducibility, and historical precedence.

The 2,2-Diphenyl-1-picrylhydrazyl ($DPPH^{\cdot}$) assay is based on the ability of antioxidants to donate a hydrogen atom or electron to neutralize the stable DPPH radical, causing a purple-to-yellow color shift. Its extended reaction time allows detection of weak or thermolabile antioxidants. However, $DPPH^{\cdot}$ is only soluble in organic solvents, reacts with other radicals, and is sensitive to light and Lewis bases, limiting its applicability in some food systems [1, 16].

The 2,2-azino-bis (3-ethylbenzothiazoline-6-sulphonic acid) ($ABTS^{\cdot+}$) assay measures antioxidant activity through the reduction of the $ABTS^{\cdot+}$ radical, leading to a color change from blue-green to pale blue. It offers greater versatility than $DPPH^{\cdot}$ due to its solubility in both aqueous and organic solvents and better stability in dark conditions. With a flexible reaction time and broad solubility, $ABTS^{\cdot+}$ is well-suited for analyzing phenolic compounds in diverse food matrices [1, 16].

The FRAP assay evaluates the reducing capacity of a sample by its ability to convert Fe^{3+} to Fe^{2+}. This redox reaction leads to an increase in absorbance, which reflects the sample's overall electron-donating potential. FRAP is simple, fast, and reproducible, though it primarily measures hydrophilic antioxidants under acidic conditions. The cupric ion reducing antioxidant capacity (CUPRAC) assay quantifies antioxidant activity based on the reduction of Cu^{2+} to Cu^{+} in the presence of chelating agents such as neocuproine or bathocuproine. The resulting Cu^{+}-ligand complex forms a stable colored product, measurable at 450 nm [1, 16].

2 Materials

2.1 Free and Bound PCs Extraction

- Ultrasonic bath.
- Rotary evaporator.
- Sodium hydroxide.
- Nitrogen gas.
- Ethanol.
- Methanol.
- Hydrochloric acid.
- Diethyl ether.
- Ethyl acetate.

2.2 Folin–Ciocalteu Assay

- Folin-Ciocalteu reagent.
- Distilled water.
- Sodium carbonate (100 g/L).
- Cuvettes (4 mL).
- Spectrophotometer.
- Ferulic acid.

2.3 DPPH$^{\cdot}$ Radical Scavenging Assay

- Methanol.
- DPPH$^{\cdot}$ (0.2 mM).
- Distilled water.
- Cuvettes (1 mL).
- Spectrophotometer.
- Trolox.

2.4 ABTS$^{\cdot+}$ Radical Scavenging Assay

- ABTS (7 mM).
- Distilled water.
- Potassium persulfate (2.45 mM).
- Microplates.
- Microplate reader.
- Trolox.

2.5 FRAP Assay

- 2,4,6-Tripyridy-s-triazine (TPTZ, 10 mM).
- Hydrochloric acid.
- Distilled water.
- Iron trichloride ($FeCl_3$, 20 mM).
- Acetate buffer (300 mM, pH 3.6).

- $FeSO_4$.
- Microplates.
- Microplate reader.

2.6 CUPRAC Assay

- $CuCl_2$.
- Ammonium acetate buffer (NH_4Ac, 1.0 M, pH 7).
- Neocuproine.
- Distilled water.

3 Methods

3.1 Free and Bound PCs Extraction

- Extract twice frozen dough or bread (4 g) with aqueous methanol solution (methanol-to-distilled water at ratio of 70:30, v/v) in an ultrasonic bath for 10 min.
- Collect the supernatants, evaporate the solvent at 40°C in rotary evaporator, and reconstitute the free PCs with 2 mL of methanol/water (70:30 v/v). Store the extract at −20°C until use.
- Digest the residues of free PCs extraction with 300 mL of NaOH 2 M at room temperature overnight by shaking under nitrogen gas. At the end of the incubation, acidify the mixtures (pH 2–3) with hydrochloric acid and extract PCs with diethyl ether/ethyl acetate (1:1 v/v). Pool the fractions and evaporated to dryness at 40°C in a rotary evaporator. Reconstitute bound PCs in 2 mL of methanol/water (70:30 v/v).

3.2 Folin–Ciocalteu Assay

- Mix 0.2 mL of PCs extract (appropriately diluted in order to achieve absorbance values within the calibration curve range) and 0.4 mL of Folin-Ciocalteu reagent (1:5 H_2O) and wait to equilibrate for 3 min.
- Add 2 mL of Na_2CO_3 (100 g/L) and allow reaction for 30 min at 30°C in darkness.
- Measure absorbance at 725 nm through a spectrophotometer.
- The amount of free or bound PCs was calculated using a ferulic acid curve and expressed as μg ferulic acid equivalent/g of dried sample.

3.3 DPPH Radical Scavenging Assay

- Mix 0.5 mL of $DPPH^{\cdot}$ solution (0.2 mM) with 0.5 mL of PCs extract (diluted as needed to keep absorbance within the calibration range).
- Allow reaction for 30 min in darkness at room temperature.
- Measures absorbance at 517 nm through a spectrophotometer.

- The DPPH· scavenging activity was calculated using a Trolox curve and expressed as μmol Trolox equivalent/g of sample.

3.4 ABTS·+ Radical Scavenging Assay (High-Throughput Methodology for Microplates)

- The ABTS·+ working reagent is prepared by mixing equal quantities of freshly prepared 7 mM ABTS solution with 2.45 mM potassium persulphate solution.
- Prepare the working ABTS·+ solution by adding 60 mL methanol into the ABTS·+ working reagent.
- Mix 198 μL of the solution with 2 μL of PCs extract (diluted as needed to keep absorbance within the calibration range) and keep the mixture at room temperature in the dark for 10 min, before reading the absorbance at 734 nm through a microplate reader.
- The ABTS·+ scavenging activity was calculated using a Trolox curve and expressed as μmol Trolox equivalent/g of dried sample.

3.5 FRAP Assay (High-Throughput Methodology for Microplates)

- Prepare the FRAP reagent solution by mixing 1 mL of TPTZ (10 mM) in 40 mM HCl, 1 mL of iron trichloride ($FeCl_3$) (20 mM), and 10 mL of acetate buffer (300 mM, pH 3.6).
- Mix 10 μL of PCs extract (diluted as needed to keep absorbance within the calibration range) with 300 μL FRAP reagent.
- Keep the mixture at room temperature for 10 min.
- Measure the absorbance at 593 nm through a microplate reader.
- $FeSO_4$ solution is used to construct the calibration curve. Results are expressed as μmol Fe^{2+}/g of dried sample.
- Trolox can also be used as positive control, and results can be expressed as μmol Trolox equivalent/g of sample, calculated from a standard curve.

3.6 CUPRAC Assay

- Prepare the CUPRAC reagent by mixing 1 mL of $CuCl_2$ solution (10 mM, in water), 1 mL of ammonium acetate buffer (NH_4Ac, 1.0 M, pH 7), and 1 mL of neocuproine solution (7.5 mM, in absolute ethanol).
- Add 1 mL each of the prepared $CuCl_2$ solution, NH_4Ac buffer, and neocuproine solution into a test tube, followed by (x) mL of PCs extract. Adjust the final volume of each sample to 4.1 mL with distilled water.
- Incubate the mixture at room temperature for 30 min.
- Measure the absorbance at 450 nm using a microplate reader.
- Trolox is used to construct the calibration curve. Results can be expressed as μmol Trolox equivalent/g of sample.

4 Notes

1. Given that the polyphenol content and antioxidant capacity of frozen dough or bread are influenced by multiple variables such as formulation, fermentation parameters, and thermal processing, it is essential to optimize both the sample mass and dilution ratio of the extracted phenolic fraction to ensure that the measured absorbance falls within the linear range of the standard calibration curve.
2. In baked products, heat-induced Maillard reaction products and endogenous proteins can interfere with spectrophotometric absorbance measurements. To minimize such interferences, sample clarification through centrifugation and/or protein precipitation is recommended. Additionally, potential background absorbance arising from browning compounds, particularly in the bread crust, should be carefully considered during data interpretation.
3. Different assays vary in sensitivity and may be interfered with. $DPPH^{\cdot}$ and FRAP are less effective for kaempferol glycosides, while $ABTS^{\cdot+}$ is ideal for colored extracts. Using multiple assays is recommended for a more accurate antioxidant assessment.
4. In the $DPPH^{\cdot}$ radical scavenging assay, the radical solution is typically prepared in methanol or ethanol, as aqueous solvents compromise radical stability. Additionally, $DPPH^{\cdot}$ is photosensitive; thus, all procedures should be conducted under reduced light or in the dark to maintain the accuracy and reproducibility of the measurements.
5. After formation, $ABTS^{\cdot+}$ should be kept in the dark for 12–16 h to ensure full $ABTS^{\cdot+}$ radical formation. Dilute to an absorbance of 0.70 ± 0.02 at 734 nm before use to ensure values fall within the calibration curve range. Freshly prepared $ABTS^{\cdot+}$ is recommended for optimal assay reliability.
6. Due to its acidic conditions (pH 3.6), the FRAP assay may degrade certain phenolics such as anthocyanins, underestimating antioxidant capacity. Use fresh extracts and minimize light/air exposure. Avoid coextraction of chelators like EDTA or phosphate that interfere with Fe^{2+}.
7. Conducted at pH 7.0 with ammonium acetate buffer, the CUPRAC assay supports phenolic stability and chromophore formation. However, browning compounds in bread crust may nonspecifically reduce Cu^{2+}. Use Trolox or quercetin as standards suitable for cereal-based samples.

References

1. Gobbetti M, Rizzello CG (2024) Basic Methods and Protocols on Sourdough. Humana, New York. ISBN 978-1-0716-3705-0. https://doi.org/10.1007/978-1-0716-3706-7
2. Perveen S, Akhtar S, Ismail T, Qamar M, Sattar D-e-shahwar, Saeed W, Younis M, Tuba Esatbeyoglu (2024) Comparison of nutritional, antioxidant, physicochemical, and rheological characteristics of whole and sprouted wheat flour. LWT 209:116679. https://doi.org/10.1016/j.lwt.2024.116679
3. Li H, Nunekpeku X, Solomon Adade SY-S, Sheng W, Kwadzokpui BA, Beloved Ahlivia E, Chen Q (2025) Phenolic compounds detection and quantification in whole grains: A comprehensive review of recent advancements in analytical methods. TrAC Trends Anal Chem 187:118215. https://doi.org/10.1016/j.trac.2025.118215
4. Heiniö RL et al (2016) Sensory characteristics of wholegrain and bran-rich cereal foods—a review. Trends Food Sci Technol 47:25–38. https://doi.org/10.1016/j.tifs.2015.11.002
5. Gaur G, Ganzle MG (2023) Conversion of (poly)phenolic compounds in food fermentations by lactic acid bacteria: novel insights into metabolic pathways and functional metabolites. Curr Res Food Sci 6:100448. https://doi.org/10.1016/j.crfs.2023.100448
6. Bertsch A, Roy D, LaPointe G (2020) Fermentation of wheat bran and whey permeate by mono-cultures of *Lacticaseibacillus rhamnosus* strains and co-culture with yeast enhances bioactive properties. Front Bioeng Biotechnol 8:956. https://doi.org/10.3389/fbioe.2020.00956
7. Zhang J, Li M, Cheng J, Zhang X, Li K, Li B, Wang C, Liu X (2021) Viscozyme L hydrolysis and Lactobacillus fermentation increase the phenolic compound content and antioxidant properties of aqueous solutions of quinoa pretreated by steaming with α-amylase. J Food Sci 86:1726–1736. https://doi.org/10.1111/1750-3841.15680
8. Xu J, Wang W, Li Y (2019) Dough properties, bread quality, and associated interactions with added phenolic compounds: a review. J FunctFoods 52:629–639. https://doi.org/10.1016/j.jff.2018.11.052
9. Han, H-M, Koh B-K (2011) Antioxidant activity of hard wheat flour, dough and bread prepared using various processes with the addition of different phenolic acids. J Sci Food Agric 91:604-608. https://doi.org/10.1002/jsfa.4188
10. Jakobek L (2015) Interactions of polyphenols with carbohydrates, lipids and proteins. Food Chem 175:556–567. https://doi.org/10.1016/j.foodchem.2014.12.013
11. Zhu F, Sakulnak R, Wang S (2016) Effect of black tea on antioxidant, textural, and sensory properties of Chinese steamed bread. Food Chem 194:1217–1223. https://doi.org/10.1016/j.foodchem.2015.08.110
12. Dhua S, Kumar K, Kumar Y, Singh L, Sharanagat VS (2021) Composition, characteristics and health promising prospects of black wheat: a review. Trends Food Sci Technol 112:780–794. https://doi.org/10.1016/j.tifs.2021.04.037
13. Xu J, Li Y, Zhao Y, Wang D, Wang W (2021) Influence of antioxidant dietary fiber on dough properties and bread qualities: a review. J Funct Foods 80. https://doi.org/10.1016/j.jff.2021.104434
14. Alara OR, Abdurahman NH, Ukaegbu CI (2021) Extraction of phenolic compounds: a review. Curr Res Food Sci 4:200–214. https://doi.org/10.1016/j.crfs.2021.03.011
15. Raposo F, Borja R, Gutiérrez-González JA (2024) A comprehensive and critical review of the unstandardized Folin-Ciocalteu assay to determine the total content of polyphenols: the conundrum of the experimental factors and method validation. Talanta 272:125771. https://doi.org/10.1016/j.talanta.2024.125771
16. Siddeeg A, AlKehayez NM, Abu-Hiamed HA, Al-Sanea EA, Al-Farga AM (2021) Mode of action and determination of antioxidant activity in the dietary sources: an overview. Saudi J Biol Sci 28:1633–1644. https://doi.org/10.1016/j.sjbs.2020.11.064

Chapter 14

Starch Content and In Vitro Simulated Digestion

Ling-Yu Qu, Run-Yang Zhang and Hua-Min Liu

Abstract

Starch, as the primary carbohydrate in bread, significantly influences its texture, shelf life, and nutritional properties. Accurate determination of starch content and its digestibility is essential for quality control and health-related evaluations. This chapter details protocols for quantifying total starch content in bread samples using enzymatic hydrolysis, followed by an in vitro simulated digestion assay to evaluate starch bioaccessibility.

Keywords Starch content, Enzymatic hydrolysis, In vitro digestion, Glucose assay

1 Introduction

Assessing both the content and digestibility of starch is essential for ensuring product consistency, prolonging shelf life, and complying with nutritional labeling and regulatory standards for starch-based products [1, 2]. The measurement of starch content across all manufacturing stages, from raw material sourcing and formulation design to processing and final quality assessment [3]. Digestibility analysis, on the other hand, is primarily conducted at specific stages of bread production, aiming to assess its nutritional profile [4], glycemic impact (e.g., GI value) [5], and to facilitate the development of functional food products [6].

Various measuring methods of total starch content have emerged in the last decades, such as acid hydrolytic methods [7], enzyme hydrolytic method, ewers polarimetric methods, and 3,5-dinitrosalicylic acid methods. Among them, enzymatic methods utilizing thermostable α-amylase and amyloglucosidase are widely accepted as accurate, reproducible, and effective approaches in analytical food science [8].

Starch digestibility was determined by two methods, including in vivo and in vitro simulations [9, 10]. Based on issues such as experimental cost and ethics, in vitro simulation is usually chosen to assess

Guoqiang Zhang and Jinying Guo (eds.), *Frozen Dough and Bread Quality*, Methods and Protocols in Food Science,
https://doi.org/10.1007/978-1-0716-5206-0_14,

the digestive properties of starch. Since starch hydrolysis follows a first-order kinetic reaction, first-order kinetic, and LOS analyses are widely applied [11]. Moreover, parameter relevant to the assessment of starch digestion characteristics is the starch hydrolysis index (HI), which is obtained by calculating the area under the digestion curve (AUC) as a percentage of the AUC of the glucose standard [12].

This chapter provides a detailed and focused introduction to the methods for determining starch content and in vitro digestion, which are currently recognized as accurate, controllable, and standardized approaches.

2 Materials

2.1 Materials for Starch Content Analysis

1. Equipment and supplies: balance (0.0001 g), thermostatic water bath (up to 100°C), reflux unit with 250 mL conical flask, small powder pulverize, electric stove, burette (25 mL), freeze dryer.
2. Chemicals and reagents: iodine, potassium iodide (KI), α -amylase (enzyme activity ≥1.6 U/mg), petroleum ether (boiling point: 60–90°C), ethyl ether, toluene, trichloromethane, hydrochloric acid, anhydrous ethanol, sodium hydroxide (NaOH), copper sulphate ($CuSO_4$), sodium potassium tartrate, potassium ferricyanide, methylene blue, methyl red, α -naphthol, sulfuric acid (concentrated, 36 N) (*see* **Note 1**).
3. D-Anhydrous Glucose, purity ≥98%.
4. Methyl red indicator: Dissolve 0.1 g methyl red in a minimal volume of 95% ethanol, then making up the solution to 50 mL with additional 95% ethanol.
5. Sodium hydroxide solution: 20 g sodium hydroxide dissolved in water and diluted to 100 mL.
6. Alkaline copper tartrate A: Dissolve 15 g $CuSO_4$ and 0.05 g methylene blue in deionized water, and dilute to a final volume of 1000 mL in a volumetric flask.
7. Alkaline copper tartrate B: Dissolve 50 g sodium potassium tartrate tetrahydrate and 75 g NaOH in deionized water. Add 4 g potassium ferricyanide and ensure complete dissolution, then dilute to 1 L in a volumetric flask (*see* **Note 2**).
8. Amylase solution: Dissolve 1 g α-amylase in 200 mL water (*see* **Note 3**).
9. Iodine solution: Dissolve 3.6 g KI in 20 mL distilled water. Introduce 1.3 g iodine into the resulting solution, ensure complete dissolution, and finally dilute to 100 mL with distilled water.

10. α-Naphthol ethanol solution: Prepare by dissolving 1 g α -naphthol in a minimal volume of 95% ethanol and making up the volume to 100 mL with additional solvent.
11. Glucose standard solution: D-Anhydrous glucose (1 g), accurately weighed after drying at 98–100°C for 2 h, is dissolved in water. Following this, 5 mL hydrochloric acid is added, and the solution is diluted to 1000 mL with water (*see* **Note 4**).

2.2 Materials for In Vitro Simulated Digestion Analysis

1. Equipment and supplies: Oven, balance, shaking water bath, spectrophotometer, centrifuge.
2. Chemicals and reagents: Dinitrosalicylic acid, sodium sulfite (Na_2SO_3), phenol, α-amylase (3700 U/g), sodium hydroxide (NaOH), pepsin (1:3000), amyloglucosidase, anhydrous ethanol.
3. D-Anhydrous Glucose, purity ≥98%.
4. DNS reagent: Prepare an aqueous solution containing 1.5 g dinitrosalicylic acid, 0.3 g phenol, 0.075 g Na_2SO_3, and 1.5 g NaOH dissolved in 150 mL deionized water.

3 Methods

3.1 Starch Content Analysis

3.1.1 Sample Preparation

1. Freeze-drying of samples.
2. Ground in a grinder and sieved (<0.425 mm).
3. Degreasing of the samples as required.

3.1.2 Enzymatic Hydrolysis Method

Processing of samples (Fig. 1)

1. Weigh 2–5 g of sample (powder) and treat with petroleum ether and 85% ethanol for degreasing and deglycosylation, respectively (*see* **Note 5**). Microglycans should also be tested, and the results should be negative (*see* **Note 6**).
2. Filter the ethanol solution dryness and transfer the retained solids to a beaker. Following a rinse of the filter paper with 50 mL water (add rinsings to beaker), heat the mixture in a boiling-water bath to complete paste formation.
3. The beaker was cooled and 20 mL of amylase solution was added when its temperature was <60°C, and the mixture is maintained at 55–60°C with stirring for 1 h.
4. No blue color should appear upon adding 0.1 mL iodine solution to 0.1 mL this solution (*see* **Note 7**).

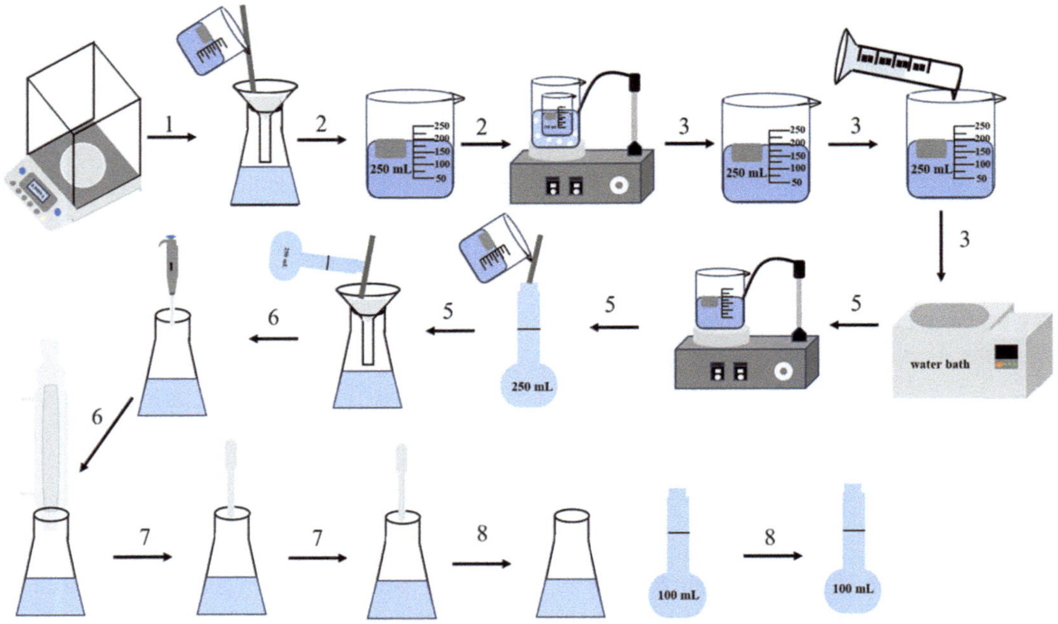

Fig.1 Schematic diagram of sample handling in starch determination

5. Heat the solution to boiling and allow to cool. Transfer the cooled solution quantitatively to a 250 mL volumetric flask. Rinse the beaker with minimal water, combining washings into the flask. Dilute to volume, homogenize, and filter (discard the initial filtrate).
6. Place 50 mL the filtrate in a 250 mL conical flask. Introduce 5 mL of 1:1 (v/v) hydrochloric acid–water mixture. Attach a reflux condenser and heat under reflux in a boiling-water bath for 60 min.
7. Upon cooling, use a disposable pipette to draw up the methyl red indicator solution and place two drops into the conical flask, and neutralize with NaOH (200 g/L).
8. Transfer the solution to a 100 mL volumetric flask. Rinse the conical flask with water and add the rinsings to the flask. Dilute to the mark with water and mix well.

Microglycans test method

1. To each test tube containing 2 mL wash solution, add α-naphthol ethanol solution (four drops, 10 g/L), followed by carefully layering 1 mL concentrated sulfuric acid down the sidewall (*see* **Note 8**).
2. A violet ring at the liquid interface indicates a positive result; a yellow-green ring indicates a negative result.

Determination of starch content (Fig. 2)

1. Using a pipette, 5 mL of basic copper tartrate A and B, 10 mL distilled water were aspirated separately and added successively to a 150 mL conical flask. In addition, two glass beads were also added. This process prepares three portions that are labeled as A_1, A_2, and A_3.
2. Add approximately 9 mL glucose standard solution dropwise from the burette to A_1, bring the solution to boiling within 2 min and maintaining boiling during the entire titration process.
3. Allow A_2 to heat to boiling within 2 min, add the sample solution dropwise from the burette and keep it boiling.
4. Add the glucose (A1) and sample (A2) solution at the rate of one drop every 2 s until the blue color of the solution in the conical flask just faded, and end this process. Register the total volume (including glucose standard solution (V1) and sample solution (V2)) consumed (*see* **Note 9**).
5. Add 1 mL less than V_2 of the sample solution to A_3 from the burette and bring to boiling within 2 min. Add the sample solution dropwise (~one drop/2 s) to the conical flask. When the color of A_3 changes from blue to colorless, end the titration process and record the volume of sample consumed ($V_{3\text{-}1}$, $V_{3\text{-}2}$, $V_{3\text{-}3}$) (*see* **Note 10**).

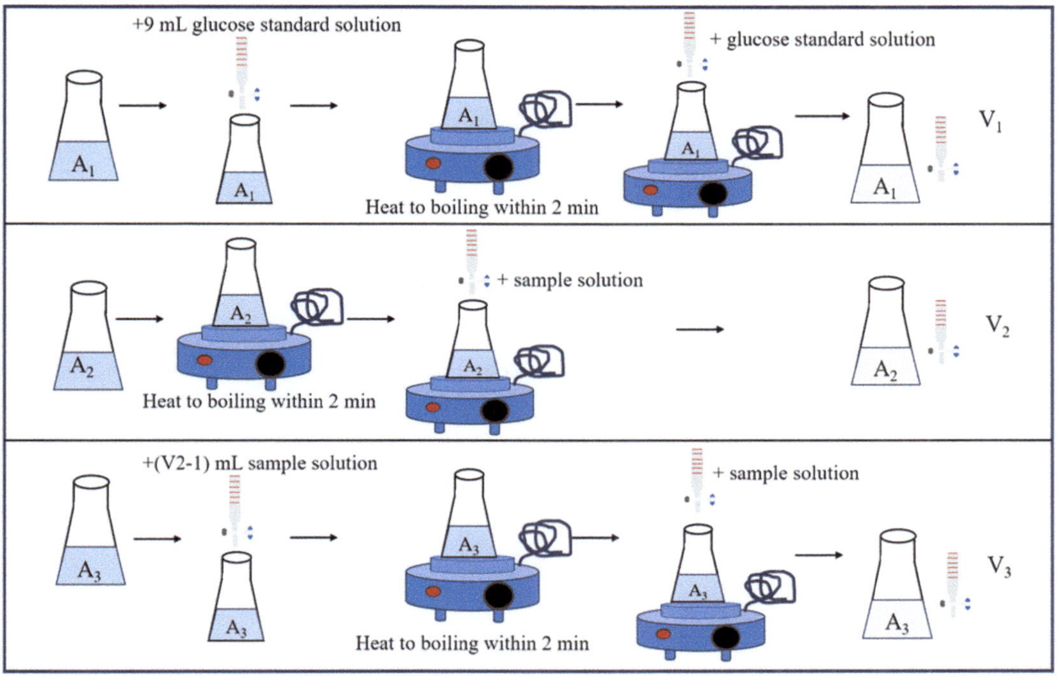

Fig. 2 Determination of starch content in sample solutions

6. Calculate the glucose equivalent mass (ml, mg) of the calibrated alkaline copper tartrate solution (1:1 mixture of A and B). The result will be calculated according to Eq. (1).

$$m_1 = \rho \times V_1 \tag{1}$$

where m_1 is the glucose equivalent mass (mg) of the calibrated alkaline copper tartrate solution (1: 1 v/v A: B). ρ is the mass concentration (mg/mL) of the glucose standard solution. V1 is the volume (mL) of glucose standard solution consumed during the calibration of the alkaline copper tartrate solution (1: 1 v/v A: B).

7. Calculate the amount of glucose in the sample solution according to Eq. (2).

$$X_1 = \frac{m_1}{\frac{50}{250} \times \frac{V_3}{100}} \tag{2}$$

where X_1 indicates the mass of glucose in the sample (mg), V_3 indicates the average volume of $V_{3\text{-}1}$, $V_{3\text{-}2}$, and $V_{3\text{-}3}$ (mL).

8. Prepare a blank by combining 20 mL water and an equivalent volume of amylase solution (as used for the sample). Titrate with the glucose standard solution. The net volume consumed (difference from the calibration volume) is recorded as V_0.

9. Calculate the amount of glucose in the reagent blank according to Eqs. (3) and (4).

$$X_0 = \frac{m_0}{\frac{50}{250} \times \frac{10}{100}} \tag{3}$$

$$m_0 = m_1 \times \left(1 - \frac{V_0}{V_1}\right) \tag{4}$$

where X_0 is the reagent blank (mg), m_0 is the Glucose content corresponding to reagent blank value (mg), V_0 is the volume of glucose standard solution consumed after addition of blank specimen (mL).

10. The starch content of the samples was calculated according to Eq. (5).

$$X = \frac{(X_1 - X_0) \times 0.9}{m \times 1000} \times 100 \tag{5}$$

where X is the starch content of the sample (g/100 g), m is the sample quality (g), 0.9 is the conversion factor for conversion of glucose to starch (*see* **Note 11**).

3.2 In Vitro Simulated Digestion Analysis

Determination of reducing sugars (DNS method)

1. Take anhydrous glucose (baked to constant weight at 80°C) and configure it into 1 mg/mL glucose standard solution.
2. Add 0, 0.2, 0.4, 0.6, 0.8, and 1.0 mL of the standard solution to a stoppered test tube, and add distilled water to 1 mL.
3. Add 2 mL of DNS reagent in a boiling water bath for 5 min, cool down as quickly as possible to room temperature.
4. Add 9 mL of distilled water.
5. Determine the absorbance value at 540 nm by spectrophotometer, and 0 mL glucose tube was used as blank control.
6. Generate the standard curve with absorbance plotted against glucose concentration.

Sample Determination (Fig. 3)

1. Weigh 0.6 g of sample and add 10 mL of 0.5% pepsin solution (*see* **Note 12**) during gastric digestion stage.
2. Shaking at 37°C was conducted for 30 min, after which 10 mL of sodium acetate buffer (pH 5.2, 0.2 mol/L) was added and the mixture was held for 10 min.
3. 5 mL of mixed enzyme solution was added to initiate digestion (porcine pancreatic α-amylase + amyloglucosidase, mixed enzyme activity ratio of 120:80 μ/mL) during intestine digestion stage.
4. At 37°C and 200 rpm shaking, the system reacted for 3 h.
5. Take 0.1 mL of sample solution at 0, 10, 20, 40, 60, 90, 120, 150, and 180 min, respectively.
6. Add 1 mL of anhydrous ethanol and vortex for 10 s to inactivate the enzyme.
7. Centrifuge at 4000 ×*g* for 5 min, take the supernatant, and then measure the reducing sugars.
8. Calculate the rate of starch hydrolysis according to the following Eq. (6):

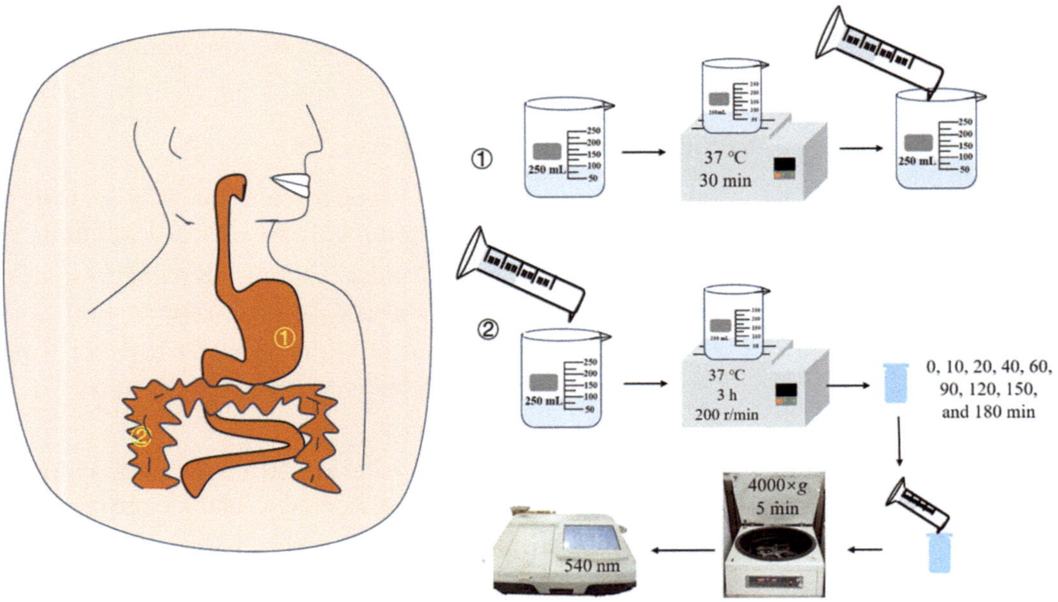

Fig.3 Schematic diagram of *in vitro* simulated digestion experiment of starch

$$\text{Hydrolysis rate}(\%) = \left(\frac{G_T \times 0.9}{TS}\right) \times 100\% \quad (6)$$

where G_T is the amount of glucose produced at T min of hydrolysis, and TS is the total starch content in the sample before hydrolysis.

9. Fitting based on first-order kinetics (Eq. 7):

$$C_T = C_\infty \times (1 - e^{-kT}) \quad (7)$$

where C_T (%) is amount of starch hydrolyzed at time T (min), C_∞ (%) is amount of starch hydrolyzed at the end of the reaction, and k (min^{-1}) is the starch digestion rate constant.

10. The slope logarithmic (LOS) equation was used to calculate k and C_∞, along with the predicted glycemic index (eGI) of the samples (Eqs. (8–11)):

$$\ln\left(\frac{dC_T}{dT}\right) = \ln(C_\infty \times k) - kT \quad (8)$$

$$AUC = \left[C_\infty \times T + \frac{C_{\infty-0}}{k} \times e^{-kT}\right]_{T_1}^{T_2} \quad (9)$$

$$HI=\frac{AUC_{sample}}{AUC_{white\ bread}} \tag{10}$$

$$eGI=0.862\times HI+8.198 \tag{11}$$

where AUC is area under the starch hydrolysis curve, and HI is the hydrolysis index.

4 Notes

1. Protective measures must be taken during the handling of sodium hydroxide and concentrated sulfuric acid, which are highly corrosive and require a high degree of vigilance.
2. The solution should be stored in a rubber stoppered glass bottle.
3. Prepare this solution fresh for immediate use. Alternatively, for storage, add a few drops of toluene as a preservative and refrigerate at 4°C.
4. The solution is then prepared when ready for use.
5. This process is carried out in a funnel with filter paper, which can be gently stirred with a glass rod to disperse the sample.
6. If the sample contained maltodextrin, it was washed with 100 mL of ethanol (85%), and then with 40% ethanol solution until the microglycans test result was negative.
7. If a blue color persists, reheat the paste, add 20 mL of amylase solution, and maintain the temperature. When the blue color disappears upon addition of iodine solution, proceed to the next step of the experimental procedure.
8. Given the ethanol-water composition of the wash solution, carefully add concentrated sulfuric acid down the tube wall. Always orient the tube mouth away from yourself and others.
9. If the predicted results in the sample solution show that its glucose is very concentration, dilution of the sample is required. This ensures the sample titration volume approximates the glucose standard volume (~10 mL) used in the alkaline copper tartrate solution calibration.
10. If the sample concentration is too low for 25 mL to reach the titration endpoint, substitute 10 mL of sample solution for the water. Subsequently, the titration process was continued

with glucose standard solution, and the assay ended when the blue color disappeared. The result was calculated according to Eqs. (12) and (13).

$$X_2 = \frac{m_2}{\frac{50}{250} \times \frac{10}{100}} \quad (12)$$

$$m_2 = m_1 \times \left(1 - \frac{V_4}{V_1}\right) \quad (13)$$

where X_2 is the mass of glucose in the sample (mg), m_2 is the mass of glucose equal to the difference between the volume of glucose standard solution consumed in calibrating the alkaline copper tartrate solution (half of solution A and half of solution B) and the volume of glucose standard solution consumed after the addition of the sample solution (mg), V_4 is the volume of glucose standard solution consumed after sample addition (mL).

11. For results <1 g/100 g, retain two significant digits. For results ≥1 g/100 g, retain three significant digits.
12. Prepare with 0.5 mol/L, pH 1.5 hydrochloric acid-KCl solution.

Acknowledgments

This work was financially supported by China Agriculture Research System of MOF, and MARA (CARS-14-1-29); Thousand Talents Program for Youth top-notch Talents of Central Plain (2023Liuhuamin).

References

1. Hsiao YT, Yang CY (2020) Microbial shelf-life, starch physicochemical properties, and in vitro digestibility of pigeon pea milk altered by high pressure processing. Molecules 25:2516
2. Corrado M, Zafeiriou P, Ahn-jarvis JH et al (2023) Impact of storage on starch digestibility and texture of a high-amylose wheat bread. Food Hydrocoll 135:108139
3. Yang ZX, Li JL, Ji ZL et al (2024) Effects of wheat starch content on its flour and frozen dough bread. Food Chem 23:10153
4. Qi KX, Yi XE, Li C (2022) Effects of endogenous macronutrients and processing conditions on starch digestibility in wheat bread. Carbohyd Polym 295:119874
5. Freitas D, Souchon I, Feunteun SL (2022) The contribution of gastric digestion of starch to the glycaemic index of breads with different composition or structure. Food Funct 13:1718–1724
6. Wahab M, Janaswamy S (2025) Porous starch granules loaded curcumin and resveratrol modulate starch digestibility of wheat bread: toward developing novel functional foods. Food Hydrocoll 162:111006
7. Hall MB (2015) Determination of dietary starch in animal feeds and pet food by an

enzymatic-colorimetric method: collaborative study. J AOAC Int 98:397–409

8. Ranathunga SD, Kalscheur KF, Anderson JL et al (2017) Evaluation of a modified method to measure total starch in animal feeds. Anim Feed Sci Tech 226:124–132
9. Rojas-Bonzi P, Vangsoe CT, Nielsen KL et al (2020) The relationship between in vitro and in vivo starch digestion kinetics of breads varying in dietary fibre. Foods 9:1337
10. Tien NNT, Anh TNQ, Phi NTL et al (2018) In vitro and in vivo starch digestibility and quality of bread substituted with acid and heat-moisture treated sweet potato starch. Starch-Starke 70:1800069
11. Edwards CH, Warren FJ, Milligan PJ et al (2014) A novel method for classifying starch digestion by modelling the amylolysis of plant foods using first-order enzyme kinetic principles. Food Funct 5:2751–2758
12. Qiao JW, Meng QX, Zhang Z et al (2024) Ferulic acid regulates proso millet starch digestibility through alteration of starch multiscale structure and inhibition of digestive enzymes. Lwt-Food Sci Technol 205:116548

Chapter 15

Determination of the Aroma Compounds

Yuanhui Wang, Yixi Yang and Guoqiang Zhang

Abstract

As one of the most globally consumed staple foods, the aroma compounds of bread play a crucial role in consumer acceptance. Sourdough have been utilized in bread production for more than 5000 years to enhance the texture and flavor of baked foods. The metabolic activities of the sourdough microbial community are exploited not only to ferment dough but also to generate complex aroma compounds that influence the sensory properties of the final baked product. Aroma constitutes a key determinant of food sensory quality and serves as a key driver of consumer acceptance. Numerous chemical reactions involving metabolites such as sugars, amino acids, and lipids occur during food processing or cooking. These chemical processes typically lead to the formation of aroma compounds. Accurate determination of aroma compounds is crucial for product quality control, flavor mechanism studies, and new product development. Gas chromatography-mass spectrometry (GC-MS) combined with solvent-assisted flavor evaporation (SAFE) has become an important analytical technique in this field. This chapter will provide a description of gas chromatography-olfactometry and aroma extract dilution analysis for the evaluation of aroma compounds in sourdough bread including sample treatment, SAFE conditions, GC-MS-O conditions, aroma extract dilution analysis, as well as quantification methods.

Keywords Bread, Aroma compounds, Gas chromatography-mass spectrometry-olfactometry (GC-MS-O), Aroma extract dilution analysis, Quantitative determination

1 Introduction

Flavor results from compounds are divided into two broad classes: taste and odor, the latter often designated as aroma substances. Aroma substances are volatile compounds that are perceived by the odor receptor sites of the smell organ, and they will reach the receptors when drawn in through the nose (orthonasal detection) and via the throat after being released by chewing (retronasal detection) [1].

Aroma is an important factor that determines the sensory quality of food, which is also the key driving force for consumers' acceptance, while aroma modification and improvement are useful and effective methods for food flavor iteration [2]. To achieve enhancement in food flavor quality, the characterization of its flavor must be conducted as the primary step. Up to now, over 10,000 aroma

Guoqiang Zhang and Jinying Guo (eds.), *Frozen Dough and Bread Quality*, Methods and Protocols in Food Science, https://doi.org/10.1007/978-1-0716-5206-0_15,

compounds—including acids, alcohols, aldehydes, ethers, amines, phenols, esters, ketones, olefins, nitrogen-containing, sulfur-containing, and heterocyclic compounds, among others—have been identified among more than 270 kinds of foods [3, 4]. Based on previous studies, aroma compounds in foods are derived mainly from three chemical processes, namely (1) enzymatic reactions, (2) microbial fermentation, and (3) chemical reactions [5–7]. The composition and concentration of aroma compounds are influenced by multiple factors including raw materials, processing methods, physical state, food matrix, storage duration, and storage conditions [2]. Therefore, the characterization and quantification of key aroma compounds become pivotal for enhancing the overall sensory quality of food products.

Currently, food aroma analysis can be broadly divided into two parts: (1) pretreatment for aroma isolation and enrichment, and (2) instrumental qualitative and quantitative analysis of aroma compounds, including the analysis of odor-active compounds.

Aroma extraction is crucial for accurate qualitative and quantitative analyses. However, some challenges occur during the separation and concentration of aroma compounds due to their small molecular weight, high volatility, low concentration, and instability, as well as due to the impacts of the complex food matrix [2]. Therefore, selecting an appropriate extraction method with high efficiency and minimal matrix interference is essential to prevent degradation of aroma compounds or formation of artifacts. The main two categories of aroma extraction methods are classified as solvent-assisted flavor evaporation (SAFE) and solvent-free aroma extraction methods, including headspace (HS) extraction, solid-phase microextraction (SPME), a mono trap (DCC18, DSC18), and thermal desorption (TD), stir-bar sorptive extraction (SBSE), and solvent aroma extraction methods, which can be categorized into three main types based on solvent extraction characteristics: solvent extraction methods (LLE, LSE, ASE), distillation methods (SDE, VDE, SAFE), and solid-phase extraction (SPE).

Solvent-assisted flavor evaporation (SAFE) is a key technology in the analysis of food aroma compounds. With its high vacuum (typically <1 Pa) and low-temperature operation, it offers significant advantages in the extraction of volatile compounds from complex matrices such as bread. Firstly, it effectively lowers the boiling point of volatile compounds, reducing degradation of heat-sensitive substances like 1-octen-3-ol caused by high temperatures, thereby maximizing the retention of the structural integrity and content of original aroma compounds. Secondly, it can suppress secondary reactions (such as the Maillard reaction) of components (sugars and amino acids), preventing the formation of nonnatural products (artifacts), and ensuring that the extracted compounds accurately reflect the intrinsic flavor of the sample. Thirdly, it separates volatile components from nonvolatile matrices such as proteins and starches, reducing interference and enhancing the detection sensitivity of GC-MS/O analysis for low-threshold compounds (e.g., (*E,E*)-2,4-decadienal). Additionally, it demonstrates excellent extraction performance for compounds with varying polarities and boiling points, is compatible with multiple detection techniques, and achieves a recovery rate of over 85% with a relative standard deviation (RSD) <5%, providing reliable and efficient technical support for flavor quality evaluation of bread and other foods.

Following the extraction and enrichment of aroma compounds from food matrices, comprehensive identification, and quantification are essential to determine the major aroma constituents and pinpoint the key odor-active compounds. Conventional aroma analysis primarily relies on gas chromatography coupled with flame ionization detection (GC-FID) and gas chromatography-mass spectrometry (GC-MS). Alternative analytical approaches, such as electronic nose (E-nose), nuclear magnetic resonance (NMR) spectroscopy, and infrared (IR) spectroscopy, are also employed. These complementary techniques exhibit broad-spectrum response profiles towards diverse aroma compounds [8].

Gas chromatography-mass spectrometry (GC-MS) efficiently separates volatile/semivolatile compounds within complex food matrices using gas chromatography (GC), and subsequently ionizes the separated components via mass spectrometry (MS). Identification is achieved by analyzing molecular weight and structural information based on the mass-to-charge ratio (m/z), supported by spectral library matching [9]. Renowned for its high sensitivity, high resolution, and strong robustness against matrix interference, GC-MS has become a core technique for food flavor analysis widely applied in the study of wheat-based products such as steamed bread and bread [10–12]. However, GC-MS has limitations in identifying trace-level odor-active compounds, and certain key odorants may remain undetected due to extremely low concentrations or weak MS response [13]. To address this gap, gas chromatography-olfactometry (GC-O) technology was developed. This system employs a split-flow design: the column effluent is divided, with one stream directed to the MS (or FID) for qualitative and quantitative analysis, and the other stream routed to an olfactometry detection port (ODP). At the ODP, a trained sensory panel assesses the odor characteristics in real-time [14]. The synergistic mechanism of "instrumental detection + human olfaction" enables GC-O to precisely capture aroma-active compounds missed by GC-MS. Owing to its unique capability in deciphering odor contribution, GC-O has evolved into a crucial tool for investigating the impact of food origin, variety, and processing techniques on aroma profiles [13].

For the detection and selection of aroma-active compounds (AACs) in food samples, the most prevalent GC-O-associated techniques currently in use are OSME (odor-specific magnitude estimation), CHARM (combined hedonic aroma response measurement), and AEDA (aroma extract dilution analysis). OSME can obtain the RI, aroma quality, and perception intensity of aroma-active compounds to determine aroma-active compounds and their significance [15]. AEDA and CHARM are both based on measuring an aroma threshold determined through a series of dilutions [16, 17].

Although detection instruments could separate and identify aroma compounds, the concentration of individual compounds cannot directly reflect their contribution to the overall aroma profile. Therefore, the odor activity value (OAV, the ratio of concentration to the threshold value) has become a key indicator for evaluating and screening key aroma compounds to achieve qualitative and quantitative analyses [18]. Before conducting the calculation, a calibration curve needs to be plotted using the concentration and response values of the standard sample of this compound, and quantitative measurements should be carried out to obtain the accurate concentration of the aroma active compounds.

This chapter will provide a description of GC-O and AEDA for the evaluation of aroma compounds in sourdough bread including sample treatment, SAFE conditions, GC-MS-O conditions, aroma extract dilution analysis, and quantification methods.

2 Material

2.1 Reagents

1. Diethyl ether anhydrous (chromatographic purity), Anhydrous sodium sulfate (analytical reagent), acetone (analytical reagent), dichloromethane (analytical reagent), 1,2-dichlorobenzene (analytical reagent), and *n*-hexane (chromatographic purity) were sourced from Innochem Technology (Beijing Innochem Technology Co., Ltd., Beijing, China).
2. 1-Pentanol (98%) were obtained from Adamas (Adamas Reagent, Ltd. Shanghai, China).
3. 1-Heptanol (99%, H0033, TCI (Shanghai) Chemical Industrial Development Co., Ltd., Shanghai, China).
4. 2-Methyl-3-heptanone (99%) was sourced from Aladdin (Aladdin Biochemical Technology Co., Ltd. Shanghai, China).
5. *n*-Alkanes standard ($C_7 \sim C_{30}$, ≥97%, 1000 μg/mL), Supelco (49451-U, Merck KgaA, Ballefonte, PA, USA).

For quantitative determination, the standard of aroma compound will be collected based on the results of qualitative determination.

2.2 Apparatus Required

1. Analytical balance.
2. A solvent-assisted flavor evaporation (SAFE) device (Glasbläserei Bahr, Germany).

3. GC-MS-O, using an Agilent 7890A GC System equipped with Agilent 5975C MSD detector.
4. A grinder (Model DLH-100, Wuyi Qiteng Electric Appliance Co., Ltd., Wuyi, China).
5. A vacuum rotary evaporator (RE-52AA, Shanghai Yarong Biochemical Instrument Factory, China).
6. A magnetic stirrer (MS-H-S, Beijing Dragon Laboratory Instruments Co., Ltd., China).
7. A gentle nitrogen stream (ST-12, Shanghai JingRuo Scientific Instrument Co., Ltd., China).
8. An Agilent 7890A GC System equipped with Agilent 5975C MSD detector (Agilent Technologies, Santa Clara, USA).
9. An olfactory detection port (ODP) (Gerstel GmbH & Co. KG, Mülheim, Germany).
10. DB-5MS capillary column (60×0.25 mm × 0.25 μm; J&W Scientific, USA).
11. DB-WAX capillary column (60×0.25 mm × 0.25 μm; J&W Scientific, USA).

3 Methods

3.1 Sample Pretreatment

1. The freshly made bread is cooled at room temperature at 25°C for 20 min, weighed and stored in the refrigerator (−20°C).
2. Bread samples stored in the refrigerator (−20°C) are first frozen in liquid nitrogen, and then ground into a fine powder using a grinder (Model DLH-100, Wuyi Qiteng Electric Appliance Co., Ltd., Wuyi, China). After grinding, the bread samples are sieved to pass through a 100 μm of sieve.

3.2 Solvent-Assisted Flavor Evaporation

1. Weigh 50 g of bread sample using analytical balance.
2. Add 500 μL of 1,2-dichlorobenzene (100 mg/L in acetone, used as the internal standard).
3. Extract with dichloromethane (200 mL) for 1 h at room temperature using a magnetic stirrer.
4. After mixing, leave for 10 min, filter the mixture solution, and collect the mixture solution.
5. Refer to Steps (3) and (4) to extract the sample residue twice more. Mix the three filtrates.
6. After mixing the resulting filtrate, concentrate filtrate to 150 mL using a vacuum rotary evaporator (RE-52AA, Shanghai Yarong Biochemical Instrument Factory, China).

7. Transfer the concentrated solution to the SAFE device (Glasbläserei Bahr, Germany, *see* **Note 1**) for distillation. The circulating water and water bath temperature are both 40°C. Add liquid nitrogen to the cold trap and liquid nitrogen basin to stabilize the vacuum degree (1×10^{-3} Pa).
8. Dry the distillate with anhydrous sodium sulfate at −20°C for 15 h.
9. Filter the extraction solution, and then concentrate the extraction solution to 5 mL at 20°C and 20 Pa.
10. Concentrate to 1 mL under a gentle nitrogen stream (ST-12, Shanghai JingRuo Scientific Instrument Co., Ltd., China).
11. The extraction operation for all samples is repeated three times, and the resulting concentrates are all stored at −18°C before analysis.

3.3 Gas Chromatography-Mass Spectrometry (GC-MS)

The concentrated extract (1 μL) is determined by GC-MS-O, using an Agilent 7890 A GC System equipped with Agilent 5975 C MSD detector (Agilent Technologies, Santa Clara, USA) and an olfactory detection port (ODP) (Gerstel GmbH & Co., KG, Mülheim, Germany). Two chromatography columns (DB-5MS and DB-WAX) of different polarity are used. The concentrated extract (1 μL) is injected directly into the GC system in splitless mode (Table 1).

3.4 Gas Chromatography-Olfactometry (GC-O)

GC-MS-O system is used for further identifying the aroma compounds. The effluent is separated in 1:1 ratio at the end of the gas chromatographic column, and two effluents are obtained in the mass selective detector and sniffing detector separately.

Table 1
Gas chromatography-mass spectrometry instrument parameter conditions

Apparatus	Parameters
DB-5MS capillary column (60 m × 0.25 mm × 0.25 μm; J&W Scientific, USA)	Carrier gas: helium (purity 99.9999%) with a flow rate of 1.0 mL/min. Injection temperature: 250°C. Transfer line temperature: 280°C. Oven temperature is programmed from an initial temperature of 40°C (2 min hold) increases to 50°C (1 min hold) at 3°C/min, then ramps to 110°C (1 min hold) at 5°C/min, finally raises to 250°C (1 min hold) at 6°C/min and up to 300°C (10 min hold).
DB-WAX capillary column (60 m × 0.25 mm × 0.25 μm; J&W Scientific, USA)	Transfer line temperature: 250°C. Other parameters are the same as the chromatographic parameters in DB-5MS program.
MS detector	Full scan mode and scan range m/z: 35-350. Electron ionization energy: 70 eV. Ion source temperature: 230°C. MS quadrupole temperature: 250°C.

Sensory panel consisted of six healthy and nonsmoking assessors, three males and three females, aged 21–25. All panelists have an acute sense of smell and taste. Sensory panelists have been professionally trained to recognize and describe the aroma of bread. Training program is according to reference and the international standards (ISO 8586: 2012) [19]. The descriptors of aroma compounds refer to the literature about bread aroma [20]. Sensory evaluators record flavor characteristics and intensity, while aroma compounds reported by two or more sensory evaluators using the same sensory descriptions are recorded in the final results. The temperature program is the same as that used in the GC-MS analysis in Sect. 3.3.

3.5 Aroma Extract Dilution Analysis (AEDA)

The flavor dilution (FD) factors of the aroma-active compounds in bread extract are analyzed by AEDA [21]. The original odor concentrate of sample extract is stepwise diluted with redistilled diethyl ether anhydrous to 1: 3^n ($n=1$–8) until no odorant is detectable by sniffing, and aliquots (1 μL) of each fraction are analyzed by GC-MS-O (*see* **Note 2**). Three trained sensory panelists (who are mentioned in Sect. 3.4) execute the AEDA test and the aroma description and intensity are recorded. Triplicate runs are performed on all samples.

3.6 Aroma Compounds Identification and Quantitation

1. ***Identification***

After identifying aroma-active compounds by GC-MS-O, the retention time (GC), mass spectrum (MS) and linear retention index (LRI) of standard compound are used to compare with those of aroma-active compounds for accurate identification.

LRIs were calculated based on a $C_7 \sim C_{30}$ n-alkane series analyzed under identical conditions using the following formula:

$$LRI = 100N + 100n\left[(t_{Ra} - t_{RN})/(t_{R(N+n)} - t_{RN})\right]$$

where N is the carbon number of the n-alkane eluting before the analyte, n is the carbon number difference between adjacent alkanes, and t_{Ra}, t_{RN}, and $t_{R(N+n)}$ are the retention times of the analyte, lower alkane, and upper alkane, respectively.

2. ***Quantitation***

The quantitative determination of aroma-active compounds is performed based on standard addition method. According to the needs of qualitative determination, all required aroma compound standard samples are purchased from chemical reagent company.

1) 2-Methyl-3-heptanone is dissolved in n-hexane as an inner standard (IS) compound to prepare the IS solution of 0.1 mg/mL (*see* **Note 3**).

2) Each aroma-active compound is accurately weighed and dissolved in *n*-hexane in a volumetric flask as standard mixed solution (1 mg/mL).
3) Dilute the standard mixed solution with *n*-hexane and add IS solution to form a series of standard solutions with 9 concentration gradients, for example, 0.1 (μg/mL), 0.5 (μg/mL), 1.0 (μg/mL), 5.0 (μg/mL), 10.0 (μg/mL), 25.0 (μg/mL), 50.0 (μg/mL), 100.0 (μg/mL), and 500.0 (μg/mL). The concentration of IS in each concentration mixed solution is constant at 0.02 mg/mL (*see* **Note 4**).
4) During the process of extracting aroma compounds from the bread using the SAFE method (*see* **Note 5**), different concentrations of the mixed standard solutions were, respectively, added to the bread matrix samples that had already had the volatiles removed. Each extract is determined by GC-MS system, and then, the calibration curve of each compound is drawn. The GC-MS analysis of each extract was executed in triplicate.

Based on the ratio of the concentration of the standard compound to the concentration of the internal standard as the x-axis and the ratio of the peak area of the standard compound to the peak area of the internal standard compound as the y-axis, the compound calibration curve is drawn, and the quantitative determination is carried out.

3.7 Method Validation

The standard deviation (SD) of aroma-active compound in the GC-MS analysis is calculated by ten replicates determination of standard solution with the minimum concentration. The limit of quantitation (LOQ) and the limit of detection (LOD) are calculated as ten times and three times the standard concentration with the minimum peak height in total ion chromatogram, respectively. The recovery rate of each determined aroma-active compound is performed by adding the standard compound with three concentrations (50%, 100%, and 200% of actual concentration) in the determination of each real sample, respectively. The sample is determined in six replicates for calculating the relative standard deviation (RSD) of each quantitative aroma-active compound.

3.8 Odor Activity Values (OAVs)

OAV is obtained by calculating the ratio of the concentration of aromatic active compounds in the sample to the olfactory threshold. The concentration of aromatic active compounds is quantified using a calibration curve, while the olfactory threshold is obtained by consulting literature. Compounds with OAV >1 are typically considered key aromatic compounds. The higher the OAV, the greater their contribution to the overall aroma of the sample. OAV calculation formula:

$$\mathrm{OAV} = \mathrm{Ci}/\mathrm{OTi}$$

In the formula, Ci represents the concentration of the compound in the sample, units μg/g; OTi represents the threshold of the compound in water, units μg/g.

4 Notes

1. During the SAFE process, solvent flow must be adequately controlled to form droplets and ensure that the solvent or aqueous solution is completely vaporized. In addition, due to the heat absorption during vaporization, the temperature of the manual valve section (40–60°C) tends to drop, especially with high-fat samples, which are prone to condensation and blockage. Furthermore, the manual valve of the SAFE equipment can result in poor yield and poses a risk of volatile separates being contaminated by nonvolatile compounds.
2. The original odor concentrate of the sample extract is diluted stepwise with redistilled diethyl ether anhydrous to 1: 3^n ($n=1$–8). The stepwise dilution being specified as follows: the first dilution was made in the ratio of 1:3. The second dilution refers to the dilution of the first diluted solution in accordance with 1:3 again, which is equivalent to obtaining a ratio of 1:9. According to this principle, the solution was further diluted, and it was diluted a total of eight times ($n=8$).
3. The internal standard (IS) must exhibit matching physicochemical properties to the target analytes. The document selected 2-methyl-3-heptanone (suitable for aldehydes) and 3-heptanol (suitable for alcohols) as the IS. However, when there is a significant difference in polarity or boiling point between the target analyte and the IS (e.g., a polar analyte like 1-pentanol vs. a nonpolar IS), the quantitative deviation can reach 15%–20%. Therefore, a compound-specific internal standard must be selected based on the chemical class of the target analyte.
4. Adding a constant amount of internal standard (IS) solution to each diluted standard solution in the beginning of preparation process is important to ensure consistent handing throughout analysis. An example of preparing 1 mL of 500.00 μg/mL of diluted mixed standard solution (containing 0.02 mg/mL of IS) from 1 mg/mL of stock mixed standard solution: 0.5 mL of 1 mg/mL of mixed standards stock solution + 0.2 mL of 0.1 mg/mL of IS solution + 0.3 mL of *n*-hexane.
5. It is important to extract at an appropriate temperature. High extraction temperatures may lead to sample decomposition or cooking. Compounds may be formed that are not present in the original food sample. Lipid oxidation, as a result of prolonged heating at high temperature, may give large aldehyde and

alcohol peaks. However, higher temperatures may be necessary when using headspace techniques to enable the extraction of low levels of higher boiling aroma compounds. Concentration of the extract should be carried out at temperatures slightly higher than the boiling point of the extraction solvent, in order to prevent losses of potentially important aroma compounds.

References

1. Belitz H-D, Grosch W, Schieberle P (2004) Aroma compounds. Food chemistry. Springer Berlin Heidelberg, Berlin, Heidelberg, pp 342–408
2. Pu D, Xu Z, Sun B et al (2025) Advances in food aroma analysis: extraction, separation, and quantification techniques. Foods 14(8):1302
3. Dunkel A, Steinhaus M, Kotthoff M et al (2014) Nature's chemical signatures in human olfaction: a foodborne perspective for future biotechnology. Angew Chem Int Ed Engl 53(28):7124–7143
4. Kaczmarska K, Taylor M, Piyasiri U et al (2021) Flavor and metabolite profiles of meat, meat substitutes, and traditional plant-based high-protein food products available in Australia. Foods 10(4):801
5. Pastorelli S, Valzacchi S, Rodriguez A et al (2006) Solid-phase microextraction method for the determination of hexanal in hazelnuts as an indicator of the interaction of active packaging materials with food aroma compounds. Food Addit Contam 23(11):1236–1241
6. Genovese A, Caporaso N, Bari VD et al (2019) Effect of olive oil phenolic compounds on the aroma release and persistence from O/W emulsion analysed in vivo by APCI-MS. Food Res Int 126:108686
7. Starowicz M, Zieliński H (2019) How maillard reaction influences sensorial properties (color, flavor and texture) of food products? Food Rev Int 35:1–19.
8. Biniecka M (2011) Analytical methods for the quantification of volatile aromatic compounds. Trends Analyt Chem 30(11):1756–1770
9. Valli SA, Sezhian U (2019) Sprouts as functional food-an approach towards the identification of natural antibiotic resistance breakers. JDDT 9:23–35
10. Majcher MA, Olszak-Ossowska D, Szudera-Kończal K et al (2020) Formation of key aroma compounds during preparation of pumpernickel bread. J Agric Food Chem 68(38):10352–10360
11. Huang Y, Wan J, Sun M et al (2023) Flavor profile disclosure of Chinese steamed breads (CSBs) by sensomics approach. Food Biosci 51:102198
12. Ji-Wei P, Fei-Long Y, Yun-Fen L, De-Zhi Q, Yuan-Li L, Dong-Di X, Ling-Yan L, Liang S (2023) Research progress of HS-SPME-GC-MS in the detection of volatile substances in fruit products. Food Mach 39(04):217–224
13. Chao LI, Runhui KE, Ming W et al (2020) Research progress on analysis of food aroma substances by GC-O/MS. Food Ferment Ind 46(2):293–298
14. Capobiango M, Mastello RB, Chin S-T et al (2015) Identification of aroma-active volatiles in banana Terra spirit using multidimensional gas chromatography with simultaneous mass spectrometry and olfactometry detection. J Chromatogr A 1388:227–235
15. Dias LG, Hacke A, Bergara SF et al (2021) Identification of volatiles and odor-active compounds of aromatic rice by OSME analysis and SPME/GC-MS. Food Res Int 142:110206
16. Al-Khalili M, Pathare PB, Rahman S et al (2025) Aroma compounds in food: analysis, characterization and flavor perception. Meas: Food 18:100220
17. Pang X, Yin H, Li J et al (2025) Molecular insights into the contribution of oak barrel aging to the aroma of beer with high alcohol content using SAFE-GC-O/AEDA and OAV calculation. Food Chem 491:145329
18. Rincón CA, Guardia AD, Couvert A et al (2019) Odor concentration (OC) prediction based on odor activity values (OAVs) during composting of solid wastes and digestates. Atmos Environ 201:1–12
19. Uselmann V, Schieberle P (2015) Decoding the combinatorial aroma code of a commercial Cognac by application of the sensomics concept and first insights into differences from a German brandy. J Agric Food Chem 63(7):1948–1956
20. Birch AN, Petersen MA, Hansen ÅS (2013) The aroma profile of wheat bread crumb influenced by yeast concentration and fermentation temperature. LWT - Food Science and Technology 50(2):480–488
21. He R-Q, Wan P, Liu J et al (2020) Characterisation of aroma-active compounds in Guilin Huaqiao white sufu and their influence on umami aftertaste and palatability of umami solution. Food Chem 321:126739

Chapter 16

Volatile Flavor Compounds Analysis

Rina Wu, Junrui Wu, Feiyu An, Wenxin Yang, Xiang Jin and Wenmin Wang

Abstract

Bread aroma, one of the most critical attributes determining consumer acceptance and product quality, is derived from complex volatile compounds formed during fermentation and baking processes. However, the composition and concentration of these volatile compounds are highly susceptible to variations in raw materials, processing parameters, and storage conditions, which may lead to inconsistent flavor profiles and quality defects such as off-flavors or rapid staling. With the growing demand for high-quality bakery products, precise characterization of volatile compounds has become essential for flavor optimization and quality standardization. Headspace solid-phase microextraction coupled with gas chromatography-mass spectrometry (HS-SPME-GC-MS) is a widely used technique for volatile compound analysis, offering high sensitivity, minimal sample preparation, and reliable quantification in complex food matrices. Therefore, this chapter describes a standardized procedure for analyzing volatile compounds in bread using HS-SPME-GC-MS. The detailed protocol is divided into three main parts, including sample preparation and handling, optimization of HS-SPME extraction parameters, and instrumental analysis coupled with data interpretation.

Keywords Bread, Volatile flavor compounds, Flavor analysis, HS-SPME-GC-MS

1 Introduction

Bread stands as one of the most widely consumed staple foods globally. The aroma profile represents an important quality attribute that enhances its commercial value and market competitiveness [1, 2]. The distinctive scent of bread arises from an intricate blend of volatile compounds generated by biochemical and thermal processes during dough fermentation and baking. These compounds encompass aldehydes, alcohols, ketones, esters, acids, pyrazines, and various other substances [3, 4]. A systematic qualitative and quantitative analysis of the flavor compounds in bread are essential to elucidate the mechanisms underlying flavor formation. Moreover, the precise control of production processes and optimization of the final flavor quality are supported by the critical insights derived from this analysis.

Various flavor compounds were formed during distinct processing stages of bread production, making significant contributions to the

Guoqiang Zhang and Jinying Guo (eds.), *Frozen Dough and Bread Quality*, Methods and Protocols in Food Science,
https://doi.org/10.1007/978-1-0716-5206-0_16,

development of its complex aroma profile [5]. During fermentation, complex microbial metabolism driven by lactic acid bacteria, yeast, and endogenous enzymes facilitates a series of enzymatic reactions involving carbohydrates, phenolic compounds, lipids, and proteins. The resulting carbohydrate metabolism, in turn, profoundly influences key bread qualities including texture, water-holding capacity, shelf life, nutritional properties, and overall flavor [6–8]. In sourdough bread, flavor and aroma substances were additionally generated by associated bacteria, as exemplified by lactic acid [9, 10]. Baking reactions were also recognized as key sources of bread aroma, with two distinct pathways being primarily involved: the Maillard reaction, characterized by complex interactions between sugars and amino acids [11]. Sugar caramelization also contributed significantly to this process [12]. Both reactions contributed to crust browning and collectively led to the formation of aroma and flavor compounds. The accurate identification and quantification of these diverse flavor compounds are essential for elucidating the mechanisms of bread flavor formation and optimizing product quality.

Quality deterioration during storage was evidenced by alterations in volatile compounds, manifested through the emergence of off-flavors or the decline of characteristic fresh bread scent [13,14]. Additionally, the formation of potentially harmful substances during high-temperature baking, particularly acrylamide, presented additional food safety considerations [15]. Consequently, the analysis of volatile compounds was considered crucial not only for flavor optimization but also for monitoring product stability and safety.

The ongoing progress in flavor chemistry, combined with rapid innovations in analytical instrumentation, has positioned HS-SPME-GC-MS as a leading method for analyzing volatile flavor substances in intricate food systems [16, 17]. HS-SPME-GC-MS combines solvent-free headspace extraction, efficient solid-phase microextraction enrichment, high-resolution GC separation, and highly sensitive MS detection, offering high analytical performance with minimal sample pretreatment. This technique has been widely applied in flavor characterization of raw materials, optimization of processing parameters, evaluation of flavor stability, and quality control of final products [18, 19]. Future integration with complementary techniques such as E-nose and GC-O could provide more comprehensive flavor analysis from both holistic and key-component perspectives.

This chapter provides a step-by-step experimental protocol for the extraction and analysis of volatile compounds in bread using HS-SPME-GC-MS, including detailed procedures for sample preparation, extraction conditions, and instrumental parameter configuration.

2 Materials

2.1 Reagents

All chemical reagents are of analytical grade, unless otherwise specified.

1. 10% (v/v) nitric acid solution (for pretreatment of glassware to remove residual metal ions).
2. Deionized water (for rinsing vessels and sample preparation).
3. Sodium chloride (NaCl, analytical grade, to be dried at 105°C to remove crystalline water).
4. Internal standards (meeting conditions such as similar chromatographic behavior and no background interference, 2-methyl-3-heptanone, 4-methyl-2-pentanone or n-octanol can be selected).
5. Methanol (analytical grade, used for preparing internal standard solutions).
6. C7–C30 normal alkane standard solution (for calculating the retention index RI).
7. Blank control reagent (consistent with the reagent used for sample preparation, for instrument residue verification).

2.2 Required instruments

All glassware should be soaked in a 10% (v/v) nitric acid solution for 24 h before use and repeatedly rinsed with deionized water.

1. High-speed cryogenic grinding machine.
2. Analytical balance.
3. Headspace sample bottle (20 mL), equipped with polytetrafluoroethylene/silicone spacer.
4. Vortex oscillator.
5. HS-SPME fiber assembly
6. Constant temperature water bath pot (temperature controlled at 60°C, used for sample balance and extraction).
7. Magnetic stirrer and stirring bar.
8. Gas chromatography-mass spectrometry (GC-MS, model reference: Agilent 7890 A GC + 5975 C MSD).
9. CP-Wax chromatographic column (Specification: 60 m×0.25 mm inner diameter, 0.50 μm film thickness).
10. Helium gas cylinders and matching pressure reducing valves.
11. Quartz liner.

3 Methods

The following section details the step-by-step protocol for the analysis of volatile compounds in bread samples. All glassware needs to be pretreated before the experiment: soaked in 10 % (v/v) nitric acid solution for 24 h to remove residual metal ions, then rinsed repeatedly with deionized water three times, and dried in a 105°C oven for later use. During the experiment, nitrile gloves and protective

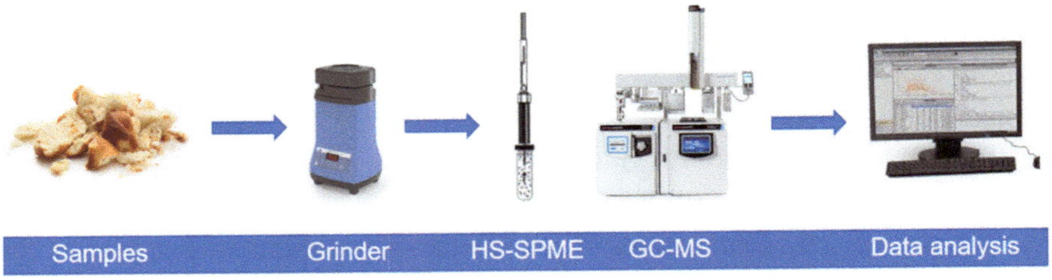

Fig. 1 Flowchart of HS-SPME/GC-MS analysis of volatile substances in bread

glasses must be worn. The operation of volatile reagents should be carried out in a fume hood. Waste should be classified and disposed of in accordance with the hazardous chemicals disposal standards. The complete analysis process is shown in Fig. 1.

3.1 Sample Preparation

1. Remove crust and retain crumb only. Cut crumb into 1 cm^3 pieces and grind into fine powder using a cryogenic grinder. Sieve through 40-mesh screen to ensure uniform particle size.
2. Precisely weigh 2.0 g of powder into a 20 mL headspace vial.
3. Add 5 mL deionized water, 1.0 g NaCl, and 10 μL internal standard solution (100 μg/mL in methanol) (*see* **Notes 1 and 2**).
4. Seal immediately with PTFE/silicone septum and cap. Vortex for 30 s to dissolve NaCl completely.

3.2 HS-SPME Extraction

1. Condition fiber in GC inlet at 250°C for 30 min prior to use.
2. Incubate sample vial at 60°C for 15 min in water bath to establish vapor-liquid equilibrium (*see* **Note 3**).
3. Expose fiber to headspace and extract at 60°C for 40 min with 500 rpm stirring (*see* **Note 4**).
4. Retract fiber and desorb in GC inlet at 250°C for 5 min.

3.3 GC-MS Analysis

1. Analyses were performed under the following GC-MS conditions (*see* **Note 5**):
   An Agilent 7890A GC coupled with a 5975C mass selective detector (MSD) was used with a CP-Wax column (60 m × 0.25 mm, 0.50 μm film thickness). The injector temperature was maintained at 250°C in splitless mode. The oven temperature was programmed as follows: 40°C for 2 min, increased to 100 °C at 2°C/min, then to 230°C at 4°C/min, and held for 3 min. Helium carrier gas was used at a constant flow rate of 0.8 mL/min. Mass spectrometry was conducted in electron impact (EI) mode at 70 eV, with a scan range of m/z 30–500. The interface, ion source, and quadrupole

Table 1
GC-MS operating parameters for volatile compounds analysis in bread

Module	Parameter category	Specification
GC	Injector temperature	250°C, splitless mode
	Oven temperature program	Hold at 40°C for 2 min, ramp to 100°C at 2°C/min, ramp to 230°C at 4°C/min, hold at 230°C for 3 min
	Carrier gas (Helium)	Constant flow rate of 0.8 mL/min
	Column	CP-Wax (60 m × 0.25 mm ID, 0.50 μm film thickness)
MS	Ionization mode	Electron Ionization (EI) at 70 eV
	Scan range	30–500 amu
	Interface temperature	280°C
	Ion source temperature	230°C
	Quadrupole temperature	150°C

temperatures were set at 280°C, 230°C, and 150°C, respectively. Detailed parameters are listed in Table 1.

2. For qualitative analysis, mass spectra were first matched against the NIST 2023 library (match factor ≥800 considered valid). Retention indices (RI) were calculated using a C7–C30 *n*-alkane standard solution and verified against literature RI values to enhance identification accuracy.
3. Quantitative analysis was performed using the internal standard method, calculating concentrations based on peak area ratios of target compounds to the internal standard.

3.4 Calculation

The content of each volatile compound in the sample was calculated using the following formula:

$$\text{Content}(\mu g/g) = \frac{As \times Cis \times V}{Ais \times Ws \times RF}$$

where,

As=peak area of target compound.
Cis=concentration of internal standard (μg/mL).
V=volume of sample solution (mL).
Ais=peak area of internal standard.
Ws=sample weight (g).

RF= relative response factor of target compound (may be omitted if RF=1 in single-point internal standard method).

4 Notes

1. Select internal standards with chromatographic behavior similar to target volatiles, no background interference, and thermal stability. Options include 2-methyl-3-heptanone, 4-methyl-2-pentanone, or *n*-octanol. Verify suitability through blank experiments.
2. Dry NaCl at 105°C for 2 h before use to remove crystalline water. Perform extraction within 30 min after sample preparation to prevent volatile loss.
3. Optimize extraction temperature based on volatile boiling points. 60°C is recommended to balance release and adsorption, while minimizing water vapor interference.
4. Determine extraction time via kinetic curves. Equilibrium is typically reached at 40 min. Maintain stirring at 500–600 rpm to avoid solution splashing.
5. For samples with high acid content, use a quartz liner in the injector port. Run blank controls every ten samples to monitor instrument contamination.

References

1. Vandevijvere S, Jaacks LM, Monteiro CA, Moubarac JC, Girling-Butcher M, Lee AC, Pan A, Bentham J, Swinburn B (2019) Global trends in ultraprocessed food and drink product sales and their association with adult body mass index trajectories. Obes Rev 20(S2):10–19
2. Bakke A, Vickers Z (2007) Consumer liking of refined and whole wheat breads. J Food Sci 72(7):473–480
3. YiNing D, Salwa K (2021) Archaeobotanical evidence reveals the origins of bread 14,400 years ago in northeastern Jordan. Compr Rev Food Sci Food Saf 20(2):1937–1981
4. Marianna R, Bartosz K, Ewa Z (2022) Effect of tomato, beetroot and carrot juice addition on physicochemical, antioxidant and texture properties of wheat bread. Antioxidants 11(11):2178
5. Birch AN, Peterson MA, Hansen AS (2014) Aroma of wheat bread crumb. Cereal Chem 91:105–114
6. Landis EA, Oliverio AM, McKenney EA, Nichols LM, Kfoury N, Biango-Daniels M, Shell LK, Madden AA, Shapiro L, Sakunala S, Kinsey D, Albert R, Matthew B, Robert RD, Noah F, Benjamin EW (2021) The diversity and function of sourdough starter microbiomes. eLife 10:e61644
7. Montanari C, Bargossi E, Lanciotti R, Chinnici F, Gardini F, Tabanelli G (2014) Effects of two different sourdoughs on the characteristics of Pandoro, a typical Italian sweet leavened baked good. LWT 59:289–299
8. Liu T, Li Y, Sadiq FA, Yang H, Gu J, Yuan L, Lee YK, He G (2018) Predominant yeasts in Chinese traditional sourdough and their influence on aroma formation in Chinese steamed bread. Food Chem 242:404–411
9. Gobbetti M (1998) The sourdough microflora: interactions of lactic acid bacteria and yeasts. Trends Food Sci Technol 9(7):267–274
10. Frasse P, Lambert S, Richard-Molard D, Chiron H (1992) The influence of fermentation on

volatile compounds in french bread dough. J Sci Food Agric 58(3):327–335
11. Romane T, Sofia N, Céline N-L, David M, Philippe J, Elodie S, Stéphanie R, Pauline MA, Céline J (2022) Effect of leavening agent on Maillard reaction and the bifidogenic effect of traditional French bread. Food Chem 393(1):133387
12. Ewa P, Sabina LW, Paulina N, Agata WB, Joanna H (2024) Enhancing antioxidant activity and modulating FODMAPs levels in inoculated sourdough rye bread. LWT 197(1):115905
13. Katsi P, Kosma I, Michailidou S, Argiriou A, Badeka A, Kontominas M (2021) Characterization of artisanal spontaneous sourdough wheat bread from central greece: evaluation of physico-chemical, microbiological, and sensory properties in relation to conventional yeast leavened wheat bread. Foods 10(3):635
14. Saeed HSO, Jing H, Xueling Z, Reham K (2023) Sorghum flour and sorghum flour enriched bread: characterizations, challenges, and potential improvements. Foods 12(23):4221
15. Andressa CL, Verônica SDB, Maristela BRC, Aline MP, Priscila TS, Eliana B-F (2024) White and wholewheat bread consumption and the risk of exposure to acrylamide and 5-hydroxymethylfurfural. Food Chem 460(3):140662
16. Pascual G-P, Pier PB, Leilei Z, Gabriele R, Luigi L (2024) Metabolomics and chemometrics: the next-generation analytical toolkit for the evaluation of food quality and authenticity. Trends Food Sci Technol 147:104481
17. Xin L, Gailing S, Jia Z, Shuang X, Lijing S, Shiming S, Jian S, Liangcai L, Cuiying Z (2025) Recent advances in analytical approaches for aroma interaction of fermented foods: a review. Food Chem 495(3):146544
18. Yanqing F, Jieqiong W, Jianxin C, Fang W, Junfeng Y, Liang Z, John S, Yongquan X (2020) Effect of baking on the flavor stability of green tea beverages. Food Chem 331(30):127258
19. Yaqian A, Ningning K, Manuel ML, Xuefeng W, Xingjiang L, Dongdong M (2024) Enhanced stability, quality and flavor of bread through sourdough fermentation with nisin-secreting *Lactococcus lactis* NZ9700. Food Biosci 62:105484

Chapter 17

Sensory Evaluation of Bread

Sijia Wang

Abstract

This chapter outlines the application of quantitative descriptive analysis (QDA) in the sensory evaluation of bread. QDA is a systematic, statistically grounded method that transforms subjective sensory experiences into quantifiable data through trained panelists and standardized terminology. It is particularly effective for assessing complex and nuanced attributes such as appearance, aroma, taste, texture, and aftertaste in bread. The method supports quality control, product development, and sensory-instrumental correlation by providing consistent, reproducible sensory profiles. The chapter also details the practical steps involved in QDA implementation, including panel selection, vocabulary development, scaling, and data analysis, emphasizing its scientific rigor and industrial relevance in bread quality evaluation.

Keywords Quantitative descriptive analysis, Bread, Sensory -evaluation, Descriptive analysis

1 Introduction

The sensory quality of bread, as a staple fermented food, is a critical determinant of its market success and consumer acceptance. The sensory properties of bread are determined not only by raw material composition, such as the protein and starch content of flour, the activity of yeast, and the proportion of salt, sugar, dairy products, and lipids, but also by technological parameters including dough mixing, fermentation conditions, proofing, and baking regimes [1, 2]. The resultant product characteristics, encompassing loaf volume, crust coloration, internal pore structure, textural parameters, aroma profiles, and flavor attributes, directly determine the market competitiveness of bread and its acceptance among consumers [3, 4]. Consumer decision-making regarding bread selection transcends mere analysis of ingredient lists; instead, it relies on integrated sensory cues, such as crust chroma and gloss, wheat-derived tanginess and toasty notes, uniformity of internal pore structure when dissected, textural composite of springiness, softness, and chewiness, and residual sweet-tart aftertaste [5]. These multifaceted sensory attributes collectively contribute to consumers' ultimate judgment and memory

Guoqiang Zhang and Jinying Guo (eds.), *Frozen Dough and Bread Quality*, Methods and Protocols in Food Science,
https://doi.org/10.1007/978-1-0716-5206-0_17,

of a particular bread. Instrumental analytical techniques alone are insufficient to fully capture or replicate this comprehensive sensory experience, so specialized sensory evaluation methods are necessary.

Food sensory evaluation is an interdisciplinary field encompassing psychology, physiology, statistics, and food science [6]. Based on research objectives and operational models, sensory evaluation is generally categorized into three main categories. Discrimination tests are employed for detecting differences, and affective tests for gauging consumer liking, but these methods offer limited diagnostic insight into the specific attributes driving perception [7, 8]. By contrast, descriptive analysis, the most systematic approach to revealing the complete sensory profile of food, involves objectively and quantitatively describing the specific sensory characteristics of a product by trained panelists [9]. In this method, quantitative descriptive analysis (QDA) has become one of the most widely used techniques due to its rigor and practicality.

QDA was developed by Tragon in the 1970s, and its implementation relies heavily on statistical analysis to establish appropriate sensory terminology, standardized procedures, and qualified panelists for product-specific evaluations—including bread [10]. QDA select appropriate terms from the vocabulary identified in simple descriptive analytical testing to quantitatively describe the entire sensory impression of a sample [11]. They can be used to evaluate a product's aroma, flavor, appearance, and texture, either individually or in combination. A hallmark of QDA is that data is not generated through consensus discussions. Instead, it uses a nonlinear scale to describe the strength of the evaluated characteristics (spiderweb plot). Changes in the plot's morphology are used to quantitatively describe changes in the sample's quality [12]. QDA is particularly effective in food sensory evaluation for quality control, quality analysis, determining differences in product properties, new product development, and product quality improvement. It provides sensory data for comparison with instrumental test data and provides a lasting record of product characteristics.

Given the complexity and subjectivity of bread's sensory profile, consumer preference tests alone are insufficient to generate systematic and accurate insights into its sensory properties. As a scientifically robust sensory analysis method, QDA addresses this gap by providing a standardized workflow for bread evaluation. It provides a standardized framework to transform subjective sensory perceptions into quantitative, actionable data [13]. Through rigorous panelist training and standardized definitions of sensory attributes, QDA transforms previously ambiguous sensory experiences into quantifiable data, enabling objective comparisons between samples [11]. For example, "This bread is soft" is translated into quantifiable data (softness score 8.2/10), allowing objective comparison between bread samples. Using multivariate statistical methods such as analysis of variance

(ANOVA) and principal component analysis (PCA), sensory profiles can be mapped, providing a solid foundation for constructing correlation models between sensory and physicochemical parameters [12] (e.g., relating moisture content to crumb softness, or reducing sugar content to sweetness). Such relationships are indispensable for optimizing formulations and processes with targeted sensory outcomes. Consequently, QDA serves as an essential tool not only for fundamental research into the factors governing bread quality but also for applied industrial practices, including quality control, shelf-life testing, and new product development, providing a scientific basis for enhancing product consistency and meeting consumer expectations.

2 Materials

2.1 Standard Sensory Analysis Laboratory

Sensory analysis laboratories should be well-ventilated, odorless, and noise-free, providing a quiet and comfortable environment for panelists. The temperature and humidity within the tasting room should be maintained at a constant level for human comfort (The room temperature is maintained at 22°C–26°C, and the humidity is between 40% and 60%). Each panelist should be provided with a relatively independent space to avoid interference. Lighting in the evaluation room should typically be vertically above the sample, and a window for passing samples and evaluation tools should be located at the front of the room.

2.2 Selection of Panelists

Experienced sensory panelists are typically selected for sensory evaluation (*see* **Note 1**). Sensory panelists who have passed the sensory analysis panelist screening test and have demonstrated a certain ability to discern differences can be considered experienced sensory panelists (*see* **Note 2**). While they can specialize in differential testing, they must regularly participate in related differential testing to maintain their ability to discern differences. Each sensory evaluation should be conducted by a team consisting atleast eight selected panelists. Team members should have the same qualifications and testing capabilities and all should be expert panelists or selected panelists (*see* **Note 3**). The team leader should be a sensory analyst.

2.3 Apparatus

The experiment requires sample plates, cups, water, pens, tissue, coded stickers, and unflavored biscuits. The sample plates holding the sample should be white, odorless, and larger than the bread being evaluated, and should not affect the evaluation results in any way. Water and unflavored biscuits are required for oral hygiene between samples. Coding stickers are used to assign a random code to each sample and are affixed to the side of the sample plates.

3 Methods

3.1 Training of Panelists

3.1.1 Determination of Evaluation Objectives

Understand the situation of related similar products, establish the best method of description and unified evaluation and identification goals, and at the same time, determine reference samples and define the vocabulary for describing characteristics.

3.1.2 Descriptive Vocabulary Development

Initial Vocabulary Generation

The panelists were asked to taste the sample and reference sample, then independently record as many words as possible to describe the bread's appearance (color, crust characteristics, etc.), smell, taste (texture, flavor, etc.), aftertaste, and etc. Each sample needs to be observed first, then smelled, and finally tasted (swallowed) (*see* **Note 4**).

Vocabulary Screening and Definition

During group discussions, share everyone's individual vocabulary and compile it into a "candidate list" (*see* **Notes 5, 6, and 7**). Duplicate words should be deleted. If you encounter words with similar meanings, discuss whether they can be merged (*see* **Note 8**). For example, baked fragrance and roasted fragrance could be changed to roasted. If the meanings are similar but subtly different, retain both words and note the difference, such as burnt fragrance and burnt bitterness. For related bread sensory evaluation terms, please refer to the bread sensory quality description in the table (Table 1) below.

3.1.3 Scale Development

Anchor Drafting

The group discusses the most suitable words as anchors. For each word, the anchor should be discussed and agreed upon by all. For example, "none" and "strong" can be used to describe the degree of sweetness, and "white" and "black" can be used to describe the degree of appearance color.

Scoring Scale Definition

A 10 cm linear scale (Line Scale) is usually used, with the left end anchored at "very weak intensity" (0) and the right end anchored at "very strong intensity" (10). The panelist marks the line and measures the distance (cm) from the mark to the left end as the intensity value of the attribute.

3.1.4 Scorecard

Based on the vocabulary, anchor points, and scoring scale discussed previously, compile a complete bread sensory evaluation scoring form for use in the formal sensory evaluation. The bread sensory

Table 1
Bread sensory quality description

Item	Soft bread	Hard bread	Puff pastry bread	Prepared bread	Other breads	Bread dry products
Shape	Complete, full, with product's proper shape	Full, multilayered, bread base complete and full, with product's proper shape	Complete, full, with product's proper shape	Complete, full, with product's proper shape	With product's proper shape	With product's proper shape
Color	With product's proper color	Bread base should have product's proper color	With product's proper color	With product's proper color	With product's proper color	With product's proper color
Texture	Delicate, elastic, even air holes	Soft interior, elastic	Fine, elastic, even air holes	Clear texture, distinct layers	With product's proper texture	With product's proper texture
Taste & Texture	With fermentation and baking aroma, soft and palatable, no off-flavor	With fermentation and baking aroma, chewy, no off-flavor	Crispy or flaky exterior, soft interior, no off-flavor	With product's proper taste and texture, no off-flavor	With product's proper taste and texture, no off-flavor	Crispy or crunchy, no off-flavor
Impurities	No visible foreign matter within normal vision range	No visible foreign matter within normal vision range	No visible foreign matter within normal vision range	No visible foreign matter within normal vision range	No visible foreign matter within normal vision range	No visible foreign matter within normal vision range

Table 2
Bread sensory evaluation scoring form

Attributes		Score
Appearance	Crust color	White ⊢———⊣ Black
	Crust thickness	Thin ⊢———⊣ Thick
	Uniformity	Irregular ⊢———⊣ Uniform
Aroma	Overall aroma intensity	None ⊢———⊣ Strong
	Milk	None ⊢———⊣ Strong
	Butter	None ⊢———⊣ Strong
	Wheat	None ⊢———⊣ Strong
	Off-flavor	None ⊢———⊣ Strong

(continued)

Table 2
(continued)

Attributes		Score
Taste	Sweetness	None ——— Strong
	Saltiness	None ——— Strong
	Sourness	None ——— Strong
	Bitterness	None ——— Strong
	Softness	Soft ——— Hard
	Moisture	None ——— Strong
	Crispness	Soft ——— Crisp
After taste	Cereal	None ——— Strong
	Oily	None ——— Strong
	Burnt	None ——— Strong

evaluation scoring form should at least include the vocabulary and scoring scale, other content can be modified based on actual circumstances. The scoring form is shown in Table 2. The complete vocabulary is based on the Descriptive Vocabulary established during the previous training.

3.2 Preparation before Sensory Test

The temperature of the sensory evaluation room was adjusted to room temperature in advance (*see* **Note 9**). Disposable plastic white plates were selected and randomly three-digit coded were affixed to the sides. Cut the bread into uniform pieces (e.g., 2 × 2 × 2 cm) and place them on white plastic plates before serving them to the panelists (*see* **Notes 10 and 11**). Each panelist was equipped with a glass of water and an unflavored biscuit in order to providing for rinsing to eliminate any aftertaste. Sensory evaluation forms and pens also need to be prepared for everyone. The evaluation time should be arranged between 9:00 and 10:00 in the morning, 3:00 and 4:00 in the afternoon, or 1 h before, or after meals.

3.3 Sensory Analysis for Bread Sample

Taste according to the corresponding blind code (*see* **Note 12**). The panelists first evaluate the appearance and smell, then touch and feel the sample. The samples are cut or torn open to inspect the internal structure. Next, panelists taste the food, assessing the mouthfeel, flavor, and other characteristics (*see* **Note 13**). Each attribute in the vocabulary is rated on a scale for intensity. The experimental process was repeated twice.

3.4 Data Collection and Analysis

The scores of all panelists for all attributes of all samples are entered into the database, and ANOVA is used to determine whether there are significant differences between different samples in each attribute

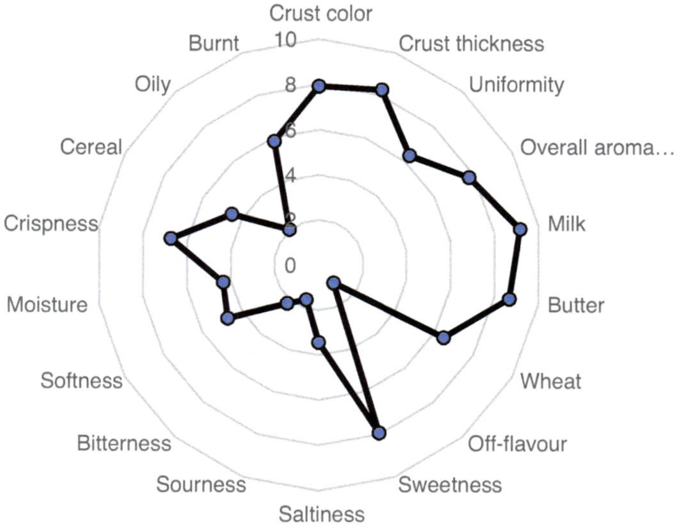

Fig. 1 Spider diagram of bread sensory evaluation

(*see* **Note 14**). The data are then displayed using spider charts/radar charts. Figure 1 is a spider diagram of bread sensory evaluation. The specific values are subject to the actual experimental results.

4 Notes

1. Panelists must be sensitive to sensory stimuli, able to accurately describe sensations, and possess an interest and commitment in the evaluation process.
2. Basic taste (sweet, sour, salty, bitter) and olfactory identification tests are used for screening.
3. Panelists are familiar with the sensory properties of various breads (e.g., whole-wheat bread, baguettes, toast, etc.) and know to use a scale for scoring.
4. Test content that needs to be recorded generally includes: (1) Defining the perceived characteristics of the bread using descriptors or related terms; (2) Evaluating the intensity of each perceived sensory characteristic of the bread; (3) Determining the presence of different sensory characteristics after swallowing (or spitting out) and the continued perception of these characteristics; (4) A comprehensive evaluation of the bread as a whole, taking into account the adaptability, intensity, and mixing of characteristics with the same background.
5. Avoid excessive word count that may lead to panel fatigue; prioritize terms that can produce statistical differences.
6. Use language that panelists can understand, using everyday sensory descriptions (avoid overly academic or technical terms unless they are familiar with them).
7. When screening vocabulary, it is necessary to ensure that each panelist clearly knows the meaning of the vocabulary and agrees on the meaning.
8. It is very important to prepare fixed reference samples for some attributes that are difficult to describe. For example, sourness reference generally uses lactic acid solution or citric acid solution with known concentration; elasticity reference generally uses a brand of toast that is recognized to have good elasticity; crispness reference generally uses Fresh baguette or soda cracker.
9. The sensory evaluation room needs to have controlled light, noise, no odor disturbance, and the temperature should be controlled at room temperature.

10. All samples should be prepared under identical conditions (same cooling time, size, and shape).
11. Panelists should avoid caffeine, tobacco, perfume, and chewing gum an hour beforehand. Rinse mouth with water between samples is necessary.
12. If there are multiple samples to be tested, they need to be presented randomly to each panelist, and no more than six samples should be presented to avoid feeling tried.
13. Rinse mouth with water and unflavored biscuits and rest for at least 1 min between each sample.
14. Before plotting, data should be checked for reproducibility (consistency between replicates by the same reviewer) and consistency (consistency between reviewers and the overall population). If individual reviewers exhibit significant deviations (systematic bias or high variability), feedback and guidance should be provided, and their data should be replaced or removed if necessary.

References

1. Mollakhalili-Meybodi N, Sheidaei Z, Khorshidian N, Nematollahi A, Khanniri E (2023) Sensory attributes of wheat bread: a review of influential factors. J Food Meas Charact 17(3):2172–2181
2. Graça C, Raymundo A, Sousa I (2019) Wheat bread with dairy products—technology, nutritional, and sensory properties. Appl Sci 9(19):4101
3. Suchintita Das R, Tiwari BK, Garcia-Vaquero M (2023) The fundamentals of bread making: the science of bread. In: Traditional European breads: an illustrative compendium of ancestral knowledge and cultural heritage. Springer International Publishing, Cham, pp 1–40
4. Dong Y, Karboune S (2021) A review of bread qualities and current strategies for bread bioprotection: flavor, sensory, rheological, and textural attributes. Compr Rev Food Sci Food Saf 20(2):1937–1981
5. Heenan SP, Dufour JP, Hamid N, Harvey W, Delahunty CM (2008) The sensory quality of fresh bread: descriptive attributes and consumer perceptions. Food Res Int 41(10):989–997
6. Lee HS (2025) Defining core science of 'sensory and consumer science': understanding human sensory perception, foods/materials interaction, and their influence on consumer behavior, health, and well-being. Food Qual Pref 129:105492
7. Rogers L, Hort J, Kemp SE, Hollowood T (2024) Discrimination testing in sensory evaluation. John Wiley & Sons, Chichester
8. Sharif MK, Butt MS, Sharif HR, Nasir M (2017) Sensory evaluation and consumer acceptability. Handb Food Sci Technol 10:362–386
9. Bleibaum RN (2020) Descriptive analysis testing for sensory evaluation. ASTM International, West Conshohocken
10. Kemp SE, Hort J, Hollowood T (eds) (2018) Descriptive analysis in sensory evaluation. Wiley-Blackwell, Chichester
11. Sidel JL, Bleibaum RN, Tao KC (2018) Quantitative descriptive analysis. In: Kemp SE, Hort J, Hollowood T (eds) Descriptive analysis in sensory evaluation. Wiley-Blackwell, Chichester, pp 287–318
12. Stone H, Sidel JL, Oliver S, Woolsey A, Singleton RC (2004) Sensory evaluation by quantitative descriptive analysis. In: Gacula MC (ed) Descriptive Sensory Analysis in Practice. Food & Nutrition Press, Trumbull, pp 23–34
13. Khatwani S (2025) Sensory evaluation techniques for food. Educohack Press, New Delhi

Subject Index

A

B

C

D

E

F

Guoqiang Zhang and Jinying Guo (eds.), *Frozen Dough and Bread Quality*, Methods and Protocols in Food Science, https://doi.org/10.1007/978-1-0716-5206-0,

T

U

V

W

Y

Z

MIX
Papier aus verantwortungsvollen Quellen
Paper from responsible sources
FSC® C105338

If you have any concerns about our products,
you can contact us on
ProductSafety@springernature.com

In case Publisher is established outside the EU,
the EU authorized representative is:
Springer Nature Customer Service Center GmbH
Europaplatz 3, 69115 Heidelberg, Germany

Printed by Libri Plureos GmbH
in Hamburg, Germany